T0076630

METHODS IN MOLECULAR BIOLOGY™

Series Editor
John M. Walker
School of Life Sciences
University of Hertfordshire
Hatfield, Hertfordshire, AL10 9AB, UK

For other titles published in this series, go to
www.springer.com/series/7651

METHODS IN MOLECULAR BIOLOGY™

Therapeutic Applications of RNAi

Methods and Protocols

Edited by

John F. Reidhaar-Olson
Cristina M. Rondinone

Hoffmann-La Roche Inc., Nutley, NJ, USA

Humana Press

Editors
John F. Reidhaar-Olson
Hoffmann-La Roche Inc.
340 Kingsland Street
Nutley NJ 07110
USA
john.reidhaar-olson@roche.com

Cristina M. Rondinone
Hoffmann-La Roche Inc.
340 Kingsland Street
Nutley NJ 07110
USA
cristina.rondinone@roche.com

ISSN 1064-3745   e-ISSN 1940-6029
ISBN 978-1-60327-294-0   e-ISBN 978-1-60327-295-7
DOI 10.1007/978-1-60327-295-7
Springer Dordrecht Heidelberg London New York

Library of Congress Control Number: 2009926661

© Humana Press, a part of Springer Science+Business Media, LLC 2009
All rights reserved. This work may not be translated or copied in whole or in part without the written permission of the publisher (Humana Press, c/o Springer Science+Business Media, LLC, 233 Spring Street, New York, NY 10013, USA), except for brief excerpts in connection with reviews or scholarly analysis. Use in connection with any form of information storage and retrieval, electronic adaptation, computer software, or by similar or dissimilar methodology now known or hereafter developed is forbidden.
The use in this publication of trade names, trademarks, service marks, and similar terms, even if they are not identified as such, is not to be taken as an expression of opinion as to whether or not they are subject to proprietary rights.
While the advice and information in this book are believed to be true and accurate at the date of going to press, neither the authors nor the editors nor the publisher can accept any legal responsibility for any errors or omissions that may be made. The publisher makes no warranty, express or implied, with respect to the material contained herein.

Printed on acid-free paper

Springer is part of Springer Science+Business Media (www.springer.com)

Preface

In the short time since its discovery, RNA interference (RNAi) has become a well-established tool in the drug discovery process. The ability to knock down expression of any gene in mammalian cells using a straightforward method has revolutionized the processes of target identification and validation, and helped to usher in the era of functional genomics. RNAi is a process in which cytoplasmic long double-stranded RNAs (dsRNAs) produced by viral infection, transposons, or introduced transgenes are targeted for inactivation. These long dsRNAs are processed into 21- to 23-nucleotide RNA duplexes by an RNase called Dicer, and are further incorporated into an RNA-induced silencing complex (RISC). The RISC uses these small RNAs to identify and cleave homologous mRNAs in the cell.

RNAi has been used to interrogate the function of candidate genes and, more recently, following the creation of random and directed siRNA libraries, has permitted phenotype-driven, reverse genetic analysis of normal physiological and disease processes. The development of stable and inducible expression vectors driving the expression of short hairpin RNAs has further expanded the application of RNAi both in tissue culture and in animal models.

Beyond its utility as a research tool, the use of RNAi as a therapeutic method promises to bring about an even greater revolution in drug discovery. The key features of RNAi – its high degree of specificity, the ubiquity of its mechanism in all cell types, its catalytic nature, and its ability to target virtually any gene in the genome – have generated excitement that RNAi-based drugs may soon become a reality. While the field of RNAi-based therapeutics is still in its early stages, research is moving ahead rapidly, with several siRNAs already being tested in the clinic.

However, significant hurdles remain, with drug delivery key among them. Delivering siRNA molecules to the right tissue and cell type to treat disease is difficult. Getting siRNAs across the cell membrane and into the cytoplasm where they can effectively silence their targets poses significant challenges. And formulating RNAi-based drugs in ways that allow convenient administration is a difficult problem as well. Currently, many delivery methods are being explored, and many have shown promising results in preclinical models. In some cases, local delivery of an unmodified siRNA may be possible with nothing more than a simple excipient as formulation. But in most cases, some sort of delivery vehicle is required. Many classes of delivery agents are under investigation, including liposomes and lipoplexes, conjugates, polymers and nanoparticles, peptides and proteins. In all of these cases, the challenge is to couple effective delivery with acceptable safety.

With few RNAi programs yet in the clinic, most of the focus remains at earlier stages: testing of delivery vehicles, identifying appropriate model systems, and evaluating the effects of RNAi in vivo. The chapters in this volume address these aspects of research toward the therapeutic application of RNAi. They describe the therapeutic applications of RNAi and potential pitfalls in oncology, viral infections, and CNS disease, using a variety of delivery methods, including liposomes, peptide-based nanoparticles, polycationic polymers, and viral vehicles. In all cases, detailed protocols are provided, to allow application of these techniques in the reader's laboratory. The collection of essays, by a team of internationally renowned authors, includes basic science chapters dealing with the biology and design of RNAi, essays describing novel strategies for delivery in vivo, and papers that discuss the application of RNAi in a variety of therapeutic areas. This volume will be of interest to basic and clinical researchers, biochemists, clinicians, molecular biologists, physiologists, and pharmacologists.

John F. Reidhaar-Olson
Cristina M. Rondinone

Contents

Contributors

PATRICK AEBISCHER • *Brain Mind Institute, Ecole Polytechnique Fédérale de Lausanne, Lausanne, Switzerland*

MORTEN ØSTERGAARD ANDERSEN • *Interdisciplinary Nanoscience Center (iNANO), Department of Molecular Biology, University of Aarhus, Aarhus C, Denmark*

FLAMINIA CATTERUCCIA • *Division of Cell and Molecular Biology, Faculty of Life Sciences, Imperial College London, London, UK*

KENNETH J. DERY • *Divisions of Molecular Biology and Immunology, Beckman Research Institute of the City of Hope, Duarte, CA, USA*

RAJESH K. GAUR • *Division of Molecular Biology and Graduate School of Biological Sciences, Beckman Research Institute of the City of Hope, Duarte, CA, USA*

SHIKHA GAUR • *Department of Clinical and Molecular Pharmacology, Beckman Research Institute of the City of Hope, Duarte, CA, USA*

MURIEL GOLZIO • *IPBS Université P Sabatier and CNRS (UMR 5089), Toulouse, France*

VERONICA GUSTI • *Division of Molecular Biology, Beckman Research Institute of the City of Hope, Duarte, CA, USA*

HEE DONG HAN • *Department of Gynecologic Oncology, The University of Texas M.D. Anderson Cancer Center, Houston, TX, USA*

KENNETH ALAN HOWARD • *Interdisciplinary Nanoscience Center (iNANO), Department of Molecular Biology, University of Aarhus, Aarhus C, Denmark*

DANIEL H. KIM • *Howard Hughes Medical Institute, Department of Molecular Biology, Massachusetts General Hospital and Department of Genetics, Harvard Medical School, Boston, MA, USA*

STEPHAN KISSLER • *Rudolf Virchow Center / DFG Center for Experimental Biomedicine, University of Würzburg, Würzburg, Germany*

JØRGEN KJEMS • *Interdisciplinary Nanoscience Center (iNANO), Department of Molecular Biology, University of Aarhus, Aarhus C, Denmark*

MOHANRAJA KUMAR • *Molecular Imaging Program, MGH/MIT/HMS Athinoula A. Martinos Center for Biomedical Imaging, Department of Radiology, Massachusetts General Hospital/Harvard Medical School, Boston, MA, USA*

ELENA A. LEVASHINA • *UPR9022 du CNRS, Group AVENIR Inserm, Institut de Biologie Moléculaire et Cellulaire, Strasbourg, France*

GABRIEL LOPEZ-BERESTEIN • *Department of Experimental Therapeutics, The University of Texas M.D. Anderson Cancer Center, Houston, TX, USA*

LINGEGOWDA S. MANGALA • *Department of Gynecologic Oncology, The University of Texas M.D. Anderson Cancer Center, Houston, TX, USA*

LAURENT MAZZOLINI • *CRPS CNRS- Pierre Fabre (UMR 2587), Toulouse, France*

ZDRAVKA MEDAROVA • *Molecular Imaging Program, MGH/MIT/HMS Athinoula A. Martinos Center for Biomedical Imaging, Department of Radiology, Massachusetts General Hospital/Harvard Medical School, Boston, MA, USA*

ANNA MOORE • *Molecular Imaging Program, MGH/MIT/HMS Athinoula A. Martinos Center for Biomedical Imaging, Department of Radiology, Massachusetts General Hospital/Harvard Medical School, Boston, MA, USA*

SHU-WING NG • *Division of Gynecologic Oncology, Department of Obstetrics, Gynecology and Reproductive Biology, Brigham and Women's Hospital/Harvard Medical School, Boston, MA, USA*

AURÉLIE PAGANIN-GIOANNI • *IPBS Université P Sabatier and CNRS (UMR 5089), Toulouse, France*

JOHN F. REIDHAAR-OLSON • *RNA Therapeutics, Hoffmann-La Roche, Inc., Nutley, NJ, USA*

CRISTINA M. RONDINONE • *Metabolic and Vascular Diseases, Hoffmann-La Roche, Inc., Nutley, NJ, USA*

JOHN J. ROSSI • *Graduate School of Biological Sciences and Division of Molecular Biology, Beckman Research Institute of the City of Hope, Duarte, CA, USA*

JOHN E. SHIVELY • *Division of Immunology, Beckman Research Institute of the City of Hope Duarte, CA, USA*

ANIL K. SOOD • *Departments of Gynecologic Oncology and Cancer Biology, The University of Texas M.D. Anderson Cancer Center, Houston, TX, USA*

JUSTIN TEISSIÉ • *IPBS Université P Sabatier and CNRS (UMR 5089), Toulouse, France*

STEPHEN MARK TOMPKINS • *Department of Infectious Diseases, College of Veterinary Medicine Center for Disease Intervention, University of Georgia, Athens, GA, USA*

CHRIS TOWNE • *Brain Mind Institute, Ecole Polytechnique Fédérale de Lausanne, Lausanne, Switzerland*

RALPH A. TRIPP • *Department of Infectious Diseases, College of Veterinary Medicine, Center for Disease Intervention, University of Georgia, Athens, GA, USA*

YUN YEN • *Department of Clinical and Molecular Pharmacology, Beckman Research Institute of the City of Hope, Duarte, CA, USA*

Chapter 1

Development and Application of a Dual-Purpose Nanoparticle Platform for Delivery and Imaging of siRNA in Tumors

Zdravka Medarova, Mohanraja Kumar, Shu-wing Ng, and Anna Moore

Abstract

The vision of using a single therapeutic agent with sufficient generality to allow application to a wide variety of diseases, yet specific enough to permit intervention at single molecular stages of the pathology, is rapidly becoming a reality through the emergence of RNA interference. RNA interference can be used to inhibit the expression of virtually any gene and, at the same time, has single-nucleotide specificity. Major challenges in applying RNA interference in vivo are adequate delivery of the siRNA molecule to the tissue of interest and methods of monitoring this delivery in a noninvasive manner. With this in mind, we have developed an approach not only to deliver siRNA to tumors, but also to track the success of the delivery by noninvasive imaging. To accomplish this, we designed a dual-function probe, MN-NIRF-siRNA, which consists of magnetic nanoparticles (MN) for magnetic resonance imaging (MRI), labeled with Cy5.5 dye for near-infrared in vivo optical imaging (NIRF), conjugated to myristoylated polyarginine peptides (MPAPs) for translocation of the complex into the cytosol, and carrying siRNA targeting tumor-specific genes. Administration of MN-NIRF-siRNA to tumor-bearing mice allowed us to monitor the delivery of the agent to tumors by MRI and NIRF imaging and resulted in efficient silencing of the target genes. This approach can significantly advance the therapeutic potential of RNA interference by providing a way not only to effectively shuttle siRNA to target sites but also to noninvasively assess the bioavailability of the siRNA molecule.

Key words: RNA interference, magnetic resonance imaging, near-infrared optical imaging, magnetic nanoparticles, myristoylated polyarginine.

1. Introduction

RNA interference is an innate cellular mechanism for post-transcriptional regulation of gene expression in which

John F. Reidhaar-Olson and Cristina M. Rondinone (eds.), *Therapeutic Applications of RNAi: Methods and Protocols, vol. 555*
© Humana Press, a part of Springer Science+Business Media, LLC 2009
DOI 10.1007/978-1-60327-295-7_1 Springerprotocols.com

double-stranded ribonucleic acid inhibits the expression of genes with complementary nucleotide sequences. Its potential for tumor therapy is indisputable, considering that one can use this mechanism to silence virtually any gene, including genes implicated in tumorigenesis, with single-nucleotide specificity *(1)*. Major obstacles in applying RNA interference in vivo are presented by the short circulation half-life of the siRNA molecule, its vulnerability to degradation by nucleases (elimination half-life 2–6 min *(2, 3)*), and the need to translocate the siRNA into the cytosol, where the RNA interference process takes place. Various approaches have been explored to overcome these obstacles. These include, but are not limited to, chemical modification, conjugation, and/or complexing of the siRNA *(2–6)*.

An essential element in the development and optimization of an siRNA delivery method is the ability to measure the bioavailability and functionality of the siRNA molecule after administration into the body. Noninvasive imaging provides the necessary set of tools to accomplish this in authentic physiologic environments and across time. The recent past has witnessed several reports describing the coupling of siRNA to contrast agents that can be used for noninvasive imaging *(4, 6–12)*.

Here, we describe the methodology behind the development and testing of a dual-purpose nanoparticle platform (MN-NIRF-siRNA) for the concurrent delivery of siRNA to tumors and assessment of the delivery by magnetic resonance imaging (MRI) and near-infrared optical imaging (NIRF). MN-NIRF-siRNA consists of a magnetic nanoparticle (MN) core, as an MRI contrast agent, labeled with Cy5.5 dye, for near-infrared in vivo optical imaging (NIRF), and conjugated to myristoylated polyarginine peptides (MPAPs) for cytosolic delivery. This complex is conjugated to an siRNA targeting model (e.g., *gfp*) or therapeutic (e.g., *birc5*) tumor-specific genes of interest. The described approach involves three major steps: (1) synthesis of the MN-NIRF-siRNA complex, (2) noninvasive imaging by MRI and NIRF imaging to assess the delivery of MN-NIRF-siRNA after in vivo administration, and (3) assessment of the silencing efficacy of MN-NIRF-siRNA.

2. Materials

2.1. Synthesis and Testing of MN-NIRF-siRNA

1. Dextran T10 (Pharmacosmos, Holbaek, Denmark).
2. $FeCl_3 \bullet 6H_2O$.
3. $FeCl_2$.
4. NH_4OH.
5. 5 M NaOH.

6. Epichlorohydrin.
7. 30% NH$_4$OH.
8. Cy5.5 NHS ester, 1 mg (GE Healthcare, Piscataway, NJ).
9. 0.5 M Sodium bicarbonate, pH 9.6.
10. G-25 Sephadex PD-10 columns (GE Healthcare, Piscataway, NJ).
11. 20 mM Sodium citrate, 0.15 M NaCl, pH 7, 7.5, 8.
12. *N*-Succinimidyl 3-(2-pyridyldithio) propionate, SPDP (Pierce Biotechnology, Rockford, IL).
13. Myristoylated polyarginine peptides (MPAPs, Myr-Ala-(Arg)$_7$-Cys-CONH$_2$).
14. 50 mM Sodium phosphate, 150 mM sodium chloride, 10 mM EDTA, pH 7.0.
15. 15 mg/mL Dithiothreitol (DTT).
16. Dimethyl sulfoxide (DMSO).
17. siRNA custom-modified with a thiol moiety via a hexyl spacer for bioconjugation (Ambion, Inc., Austin, TX).
18. 3% Tris-(2-carboxyethyl)phosphine hydrochloride (TCEP) (Sigma, St. Louis, MO).
19. 3-Maleimidobenzoic acid *N*-hydroxysuccinimide ester (MBS) (Sigma, St. Louis, MO).
20. Quick Spin Column G-50 Sephadex column (Roche Applied Science, Indianapolis, IN).
21. QuixStand benchtop filtration system with a 0.1-μm cartridge and a 100-kDa cartridge (GE Healthcare, Piscataway, NJ).
22. Bruker MQ20 Minispec NMR spectrometer (Bruker Biospin Co., Billerica, MA).
23. Sub-micron particle size analyzer (Coulter N-4, Hialeah, FL).
24. Standard tube shaker/rotator.
25. Standard spectrophotometer.
26. Standard benchtop centrifuge.

2.2. Imaging the Delivery of MN-NIRF-siRNA to Tumors

1. Five- to six-week-old female nu/nu mice.
2. Tuberculin syringes, 1 cc (Becton Dickinson, Franklin Lakes, NJ).
3. Tumor cell line of interest.
4. Animal anesthesia system Isotec 4 (Surgivet/Anesco, Waukesha, WI).
5. 9.4 T GE magnet with Bruker Biospin Avance console equipped with ParaVision 3.0.1 software and a 3 × 4 cm elliptical surface coil to transmit and receive (Bruker BioSpin, Billerica, MA).
6. Whole-body optical imaging system (Imaging Station IS2000MM, Kodak, New Haven, CT), equipped with a band-pass filter at 630 nm and a long-pass filter at 700 nm (Chroma Technology Corporation, Rockingham, VT).

2.3. Assessment of Target Gene Silencing

1. Animal anesthesia system Isotec 4 (Surgivet/Anesco, Waukesha, WI).

2. Whole-body optical imaging system (Imaging Station IS2000MM, Kodak, New Haven, CT), equipped with the following filters (Chroma Technology Corporation, Rockingham, VT):
   - for imaging of GFP: excitation, 465 nm; emission, 535 nm.
   - for imaging of RFP: excitation, 535 nm; emission, 600 nm.

3. Sodium pentobarbital injected intraperitoneally (200 mg/kg, intraperitoneal) (Sigma, St. Louis, MO).

4. Standard surgery kit (scissors, forceps) (Roboz Surgical Instrument Co., Inc., Gaithersburg, MD).

3. Methods

The synthesis of MN-NIRF siRNA consists of four distinct steps: synthesis of MN; conjugation of the fluorescent dye Cy5.5 to MN; conjugation of the membrane translocation peptide, MPAP, to MN(Cy5.5); and conjugation of siRNA to MN(Cy5.5)(MPAP). These steps are outlined in **Fig. 1.1** and described in **Sections 3.1–3.4**.

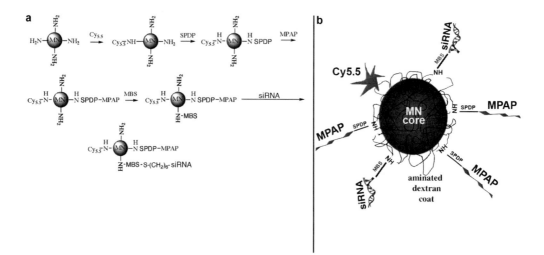

Fig. 1.1. (a) Step-by-step synthesis of the MN-NIRF-siRNA probe, by the sequential conjugation of three different entities onto magnetic nanoparticles. (b) The resultant probe consists of magnetic nanoparticles labeled with near-infrared Cy5.5 dye (NIRF) and linked through two different linkers to membrane translocation peptides (MPAPs) and siRNA molecules. (Reproduced with minor changes from Medarova et al. (6), with permission from Nature Publishing Group).

Since MN-NIRF-siRNA carries both an MRI component (MN) and a near-infrared fluorescent dye (Cy5.5), the delivery of the agent is assessed by both MRI and NIRF imaging (**Sections 3.5–3.8**).

In proof-of-principle experiments, the target gene may encode a fluorescent protein (GFP, RFP), in which case silencing can be assessed by noninvasive optical imaging or by other routine fluorescence assays, e.g., flow cytometry and fluorescence microscopy. When the target gene does not encode a fluorescent protein, silencing cannot be directly assessed by noninvasive optical imaging. Instead, alternative assays, such as quantitative RT-PCR (qRT-PCR), Western blotting, or immunohistology, can be used. In either case, validation of the silencing effect by qRT-PCR is essential. In **Sections 3.9** and **3.10**, we are restricting ourselves to a description of noninvasive optical imaging for the assessment of target gene silencing, since the other methods can vary with application and their implementation is routine.

3.1. Synthesis of MN (See Note 1)

The synthesis of dextran-coated MN has been described previously *(13)*.

1. A desired amount of dextran T10 is dissolved in hot (70°C) water and incubated for 2 h. At the end of the incubation, $FeCl_3 \bullet 6H_2O$ is added to the dextran solution and the resultant mixture is cooled down to 2–4°C.
2. Freshly prepared $FeCl_2$ in H_2O is added to the dextran–$FeCl_3$ solution, followed by neutralization with NH_4OH.
3. The resultant suspension is heated to 80°C over 1 h and then incubated at this temperature for 75 min.
4. The suspension is then passed through the QuixStand filtration system to remove large particles, free dextran, ionic iron, NH_4Cl, and NH_4OH using first a 0.1-μm cartridge and then a 100-kDa cartridge.
5. Five volumes of 5 M NaOH and two volumes of epichlorohydrin are added with constant stirring, followed by incubation for 8 h. This is done under a fume hood.
6. Eight and a half volumes of 30% NH_4OH is added followed by an overnight incubation at room temperature with vigorous stirring.
7. Free epichlorohydrin and NH_4OH are removed by ultrafiltration on QuixStand with a 100-kDa cartridge, and six washes with 20 mM sodium citrate, 0.15 M NaCl, pH 8 (8–12 l).
8. The resulting colloid is tested for free dextran (caramelization), material concentration, particle size by light scattering using a submicrometer particle size analyzer, and magnetic properties by using a Minispec NMR spectrometer.

3.2. Conjugation of Cy5.5-NHS Ester to MN

1. The pH of MN is adjusted to 9.6 with 0.5 M sodium bicarbonate.
2. The product (~10 mg Fe) is added to 1 mg of Cy5.5-NHS ester dye, followed by incubation on a rotator overnight at room temperature.
3. The mixture is purified from nonreacted dye on a Sephadex G-25, PD-10 column equilibrated with 20 mM sodium citrate buffer, 0.15 M NaCl, pH 7.5.

3.3. Conjugation of MPAP to MN-(Cy5.5)

The synthesis of the myristoylated polyarginine membrane translocation module (MPAP) has been described previously *(14)*. MPAP consists of a hydrophobic 14-carbon moiety of myristic acid, in combination with a hepta-arginine peptide. For conjugation of MPAP to MN-(Cy5.5):

1. MN-(Cy5.5) is conjugated to a heterobifunctional crosslinker *N*-succinimidyl 3-(2-pyridyldithio) propionate, SPDP, via the *N*-hydroxy succinimide ester, by co-incubation of an excess of SPDP, dissolved in DMSO, with MN-(Cy5.5) for 15 min at room temperature.
2. The conjugation is followed by purification using a Sephadex G-25, PD-10 column equilibrated with 20 mM sodium citrate and 0.15 M sodium chloride, pH 8.0.
3. The delivery module, MPAP (Myr-Ala-$(Arg)_7$-Cys-$CONH_2$), is then coupled to the linker via a sulfhydryl-reactive pyridyl disulfide residue in 20 mM sodium citrate and 0.15 M sodium chloride, pH 8.0 (*see* **Note 2**).
4. The double-labeled MN-(Cy5.5)(MPAP) is then purified by size exclusion chromatography using a Sephadex G-25, PD-10 column using a buffer containing 20 mM sodium citrate and 0.15 M sodium chloride, pH 7.0.

3.4. Conjugation of siRNA to MN-(Cy5.5)(MPAP)

1. MN-(Cy5.5)(MPAP) is coupled to an excess of MBS crosslinker.
2. The intermediate is purified using a Sephadex G-25, PD-10 column using a buffer containing 50 mM sodium phosphate, 150 mM sodium chloride, 10 mM EDTA, pH 7.0.
3. The custom-synthesized siRNA duplex is provided by the manufacturer (Ambion, Austin, TX), modified with a thiol moiety via a hexyl spacer for bioconjugation. Prior to labeling, the disulfide protecting group on $5'$-S-S-$(CH_2)_6$ is deprotected using 3% tris-(2-carboxyethyl)phosphine hydrochloride (TCEP).
4. The free thiol-siRNA is then allowed to react with MN-(Cy5.5)(MPAP) via the MBS crosslinker in 50 mM sodium phosphate, 150 mM sodium chloride, 10 mM EDTA, pH 7.0 at 4°C for 1 h.

5. The product is then purified using a Quick Spin Column G-50 Sephadex column (Roche Applied Science, Indianapolis, IN).

6. The labeling ratio of Cy5.5 per MN crystal is determined spectrophotometrically as the number of Cy5.5 dye molecules attached to a single particle. The dye-to-particle ratio is obtained from concentrations of Cy5.5 and iron. Iron concentration is determined spectrophotometrically *(15)*. For the Cy5.5 dye, the number of dyes per particle is obtained from absorption at 678 nm and an extinction coefficient of 250,000 M^{-1} cm^{-1}. The labeling of MPAP per crystal is determined based on the release of pyridine-2-thione at 343 nm ($\varepsilon = 8.08 \times 10^3$ $M^{-1}cm^{-1}$) after the addition of the reducing agent, TCEP (35 mM in DMSO).

7. The R1 and R2 relaxivities of the sample are determined at 37°C using a Minispec NMR spectrometer. These relaxivity values are measures of the longitudinal (R1) and transverse (R2) magnetization properties of a sample and are represented as $mmol^{-1}$ s^{-1} for a given temperature.

8. Nanoparticle size is measured by light scattering.

3.5. Tumor Model

1. Five- to six-week-old female nu/nu mice are injected subcutaneously with a tumor cell line of interest ($\sim 3 \times 10^6$ tumor cells). The cell line may be stably transformed with a gene encoding a fluorescent protein (GFP, RFP), in which case target gene silencing can be assessed using noninvasive optical imaging. Animals are imaged on days 10–14 after inoculation, when the tumors are $\sim$0.5 cm in diameter.

3.6. Magnetic Resonance Imaging

MRI is performed before as well as 24 h after injection of MN-NIRF-siRNA (10 mg Fe/kg; $\sim$440 nmol siRNA/kg) into the tail vein of the mouse (*see* **Note 3**).

1. For imaging, the mouse is anesthetized using isofluorane, 1.25% in 33% O_2, balance N_2O and placed prone in the magnet (9.4 T GE magnet with a Bruker Biospin Avance console equipped with ParaVision 3.0.1 software).

2. A 3 × 4 cm elliptical surface coil is used to transmit and receive.

3. The mouse is imaged using the following protocols:
   a) Spin echo axial T2-weighted imaging. Parameters: TR/TE = 6000/60.49 ms, FOV = 3.2 × 3.2 cm^2, matrix size 128 × 128, resolution 250 × 250 μm^2, slice thickness = 0.5 mm (*see* **Note 4**).
   b) MSME axial T2-weighted imaging (T2 map). Parameters: 3000/8, 16, 24, 32, 40, 48, 56, 64 ms; FOV = 3.2×3.2 cm^2, matrix size 128×128, resolution 250× 250 μm^2, and slice thickness = 0.5 mm (*see* **Note 5**).

4. After the pre-contrast (before injection of MN-NIRF-siRNA) and post-contrast (after injection of MN-NIRF-siRNA) images have been acquired, the data are analyzed using Marevisi 3.5 software (Institute for Biodiagnostics, National Research Council, Canada): MSME T2-weighted images are analyzed on a voxel-by-voxel basis by fitting the T2 measurements from the eight echo times (TE) to a standard exponential decay curve, defined by the formula $y = A^*\exp(-t/T2)$. Examples of MR images and T2 relaxation times are shown in **Fig. 1.2a**.

5. The tumors and adjacent muscle tissue are manually segmented out on these images. Their T2 relaxation times (ms) are computed by averaging the T2 relaxation times of the voxels within the tumor/muscle region of interest (ROI) from all of the slices incorporating tumor/muscle tissue (*see* **Note 6**).

3.7. In Vivo Near-Infrared Optical Imaging

Near-infrared optical imaging is performed immediately after each MRI session.

1. The mouse is anesthetized using isofluorane, 1.25% in 33% O_2, balance N_2O.

2. The animal is placed supine into a whole-body optical imaging system (Imaging Station IS2000MM, Kodak Scientific Imaging System, New Haven, CT, *see* **Note 7**).

3. For in vivo imaging, the following parameters are used: exposure time, 30.05 s; F-stop, 2.8; FOV, 100 mm; resolution, 260 dpi.

4. Examples of NIRF images are shown in **Fig. 1.2b**.

3.8. Ex Vivo Near-Infrared Optical Imaging

1. The animals are sacrificed with a high dose of sodium pentobarbital injected intraperitoneally (200 mg/kg IP).

2. The tumors and adjacent muscle tissue are excised and placed in the optical imaging system.

3. Tumor and muscle tissue are imaged using the following imaging parameters: exposure time, 30.05 s; F-stop, 2.8; FOV, 36.67 mm; resolution, 856 dpi.

3.9. Assessment of Target Gene Silencing: In Vivo Fluorescence Optical Imaging

1. Time-course in vivo fluorescence optical imaging is performed first at the same time as NIRF imaging, immediately after the pre-contrast MRI session, and after that every 12–24 h for a time period defined by the half-life of the target protein (*see* **Note 8**).

2. The mouse is anesthetized using isofluorane, 1.25% in 33% O_2, balance N_2O.

3. The animal is placed supine into a whole-body optical imaging system (Imaging Station IS2000MM, Kodak Scientific Imaging System, New Haven, CT).

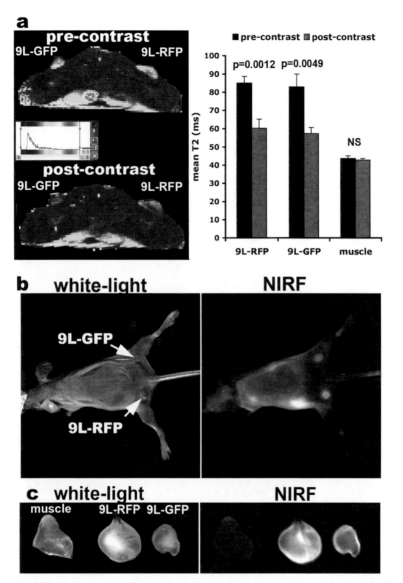

Fig. 1.2. (**a**) In vivo MRI of mice bearing bilateral 9L rat gliosarcoma tumors stably transfected with either GFP or RFP before and 24 h after MN-NIRF-siGFP administration. After injection of the probe, there is a significant drop in tumor T2 relaxation times, whereas the T2 relaxation times of muscle tissue remain unchanged. (**b**) In vivo NIRF optical imaging of the same mice as in (a) produces a high-intensity NIRF signal associated with the tumors. (**c**) Ex vivo NIRF optical imaging demonstrates a higher fluorescence in tumors than in muscle tissue. This confirms the delivery of the MN-NIRF-siGFP probe to these tissues. (Reproduced from Medarova et al. (*6*), with permission from Nature Publishing Group).

4. For in vivo imaging, the following parameters are used: exposure time, 30.05 s; F-stop, 2.8; FOV, 100 mm; resolution, 260 dpi. For imaging of GFP, the following filters are used: excitation, 465 nm; emission, 535 nm. For imaging of RFP, the following filters are used: excitation, 535 nm; emission, 600 nm (*see* **Notes 9** and **10**).

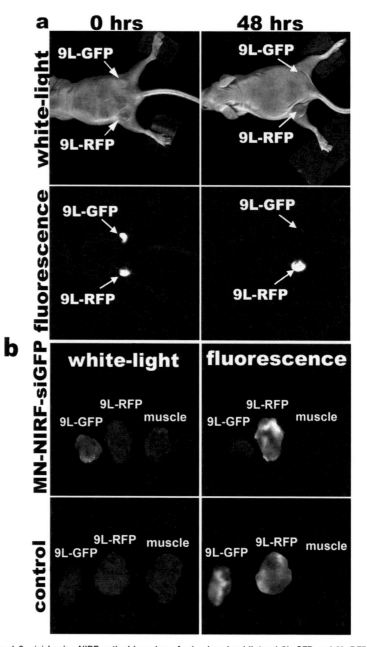

Fig. 1.3. **(a)** In vivo NIRF optical imaging of mice bearing bilateral 9L-GFP and 9L-RFP tumors 48 h after intravenous probe injection. There is a marked decrease in 9L-GFP-associated fluorescence and no change in 9L-RFP fluorescence. To generate GFP/RFP reconstructions, GFP and RFP images are acquired separately and then merged. **(b)** Correlative ex vivo fluorescence optical imaging shows a drop in fluorescence intensity in 9L-GFP tumors. There is no evidence of silencing in saline-injected controls. (Reproduced from Medarova et al. *(6)*, with permission from Nature Publishing Group).

5. Examples of in vivo fluorescence optical images are shown in **Fig. 1.3a**.

3.10. Assessment of Target Gene Silencing: Ex Vivo Fluorescence Optical Imaging

1. Ex vivo fluorescence optical imaging is performed at the same time as NIRF imaging.
   Tumor and muscle tissue are imaged using the following imaging parameters: exposure time, 30.05 s; F-stop, 2.8; FOV, 36.67 mm; resolution, 856 dpi. Filters for GFP and RFP are used, as specified above.
2. Examples of ex vivo fluorescence optical images are shown in **Fig. 1.3b**.

4. Notes

1. Considering that the MN-NIRF-siRNA agent incorporates ribonucleic acid molecules, the entire synthesis has to be completed under RNase-free conditions. All buffers and glassware need to be autoclaved or treated with RNaseZap® (Ambion, Austin, TX). Only sterile barrier pipette tips need to be used.
2. The advantage of this linker is the affordable chromophore of pyridine-2-thione, which was released after the sulfhydryl exchange between the cysteine side chain on MPAP and 2-pyridyl-disulfide group.
3. A dose of 10 mg Fe/kg is a standard dose of iron oxide contrast agent used in animal MRI studies. The amount of siRNA delivered depends on the coupling ratio of siRNA to MN achieved during synthesis. In our experience, a ratio of five siRNA molecules per MN crystal is sufficient for imaging and delivery.
4. The spin echo sequence is used for initial visualization of the ROI within the context of T2 relaxation.
5. The MSME sequence allows accurate measurement of the T2 relaxation time parameters of the tissues of interest.
6. Muscle tissue is used to define a reference T2 relaxation time to which to compare tumor relaxation time in order to determine if there is tumor-selective accumulation of the contrast agent.
7. The IS2000MM Kodak Scientific Imaging System is equipped with a 150-watt high-intensity halogen illuminator, which emits broadband white light. As recommended by the supplier (Eastman Kodak Company, New Haven, CT) for detection of Cy5.5 fluorescence, an optical band-pass excitation filter (X625, Eastman Kodak Company, New Haven, CT) is used to supply photons at 630 nm,

with an attenuation of 0.5 OD at 700 nm, which is the wavelength characterizing the long-pass emission filter (e700WA, Eastman Kodak Company, New Haven, CT). Emitted light is collected using a thermoelectrically cooled CCD camera.

8. Time-course optical imaging is necessary in order to identify the time point associated with the maximum silencing as well as the longevity of the silencing effect. This is determined by the balance between the half-life of the target protein and the proliferation rate of the target cell. For rapidly dividing tumor cells stably expressing the firefly luciferase protein, for instance, the silencing effect typically persists for 7–10 days, whereas for non-dividing cells, this time period can extend for 3–4 weeks *(16)*. For GFP/EGFP, whose half-life is about 26 h *(17, 18)*, efficient silencing (70–95%) has been reported as early as 24–48 h after transfection (http://www.ambion.com/techlib/spec/sp_4626.pdf).

9. Native tissue is associated with high absorption of visible light (~400–650 nm), reducing the sensitivity of detection of in vivo imaging studies within this range of wavelengths. Therefore, it is essential that the transgene is expressed by tumor cells at high enough levels for in vivo detection.

10. To generate GFP/RFP reconstructions based on the in vivo or ex vivo fluorescence optical images, GFP and RFP images are acquired separately without repositioning the animal and then merged using the Kodak 1DTM 3.6.3 Network software.

Acknowledgments

The authors would like to acknowledge Pamela Pantazopoulos (Martinos Center for Biomedical Imaging, MGH) for excellent technical support. This work was partially supported by K99 CA129070 to Z.M.

References

1. Brummelkamp, T. R., Bernards, R., and Agami, R. (2002) Stable suppression of tumorigenicity by virus-mediated RNA interference. *Cancer Cell* **2**, 243–247.
2. Soutschek, J., Akinc, A., Bramlage, B., Charisse, K., Constien, R., Donoghue, M., Elbashir, S., Geick, A., Hadwiger, P., Harborth, J., John, M., Kesavan, V., Lavine, G., Pandey, R. K., Racie, T., Rajeev, K. G., Rohl, I., Toudjarska, I., Wang, G., Wuschko, S., Bumcrot, D., Koteliansky, V., Limmer, S., Manoharan, M., and Vornlocher, H. P. (2004) Therapeutic silencing of an endogenous gene by systemic administration of modified siRNAs. *Nature* **432**, 173–178.
3. Morrissey, D. V., Lockridge, J. A., Shaw, L., Blanchard, K., Jensen, K., Breen, W., Hartsough, K., Machemer, L., Radka, S.,

Jadhav, V., Vaish, N., Zinnen, S., Vargeese, C., Bowman, K., Shaffer, C. S., Jeffs, L. B., Judge, A., MacLachlan, I., and Polisky, B. (2005) Potent and persistent in vivo anti-HBV activity of chemically modified siRNAs. *Nat Biotechnol* **23**, 1002–1007.

4. Bartlett, D. W., Su, H., Hildebrandt, I. J., Weber, W. A., and Davis, M. E. (2007) Impact of tumor-specific targeting on the biodistribution and efficacy of siRNA nanoparticles measured by multimodality in vivo imaging. *Proc Natl Acad Sci USA* **104**, 15549–15554.

5. Kumar, P., Wu, H., McBride, J. L., Jung, K. E., Kim, M. H., Davidson, B. L., Lee, S. K., Shankar, P., and Manjunath, N. (2007) Transvascular delivery of small interfering RNA to the central nervous system. *Nature* **448**, 39–43.

6. Medarova, Z., Pham, W., Farrar, C., Petkova, V., and Moore, A. (2007) In vivo imaging of siRNA delivery and silencing in tumors. *Nat Med* **13**, 372–377.

7. Chen, A. A., Derfus, A. M., Khetani, S. R., and Bhatia, S. N. (2005) Quantum dots to monitor RNAi delivery and improve gene silencing. *Nucleic Acids Res* **33**, e190.

8. Derfus, A. M., Chen, A. A., Min, D. H., Ruoslahti, E., and Bhatia, S. N. (2007) Targeted quantum dot conjugates for siRNA delivery. *Bioconjug Chem* **18**, 1391–1396.

9. Liu, N., Ding, H., Vanderheyden, J. L., Zhu, Z., and Zhang, Y. (2007) Radiolabeling small RNA with technetium-99m for visualizing cellular delivery and mouse biodistribution. *Nucl Med Biol* **34**, 399–404.

10. Chang, E., Zhu, M. Q., and Drezek, R. (2007) Novel siRNA-based molecular beacons for dual imaging and therapy. *Biotechnol J* **2**, 422–425.

11. Bakalova, R., Zhelev, Z., Ohba, H., and Baba, Y. (2005) Quantum dot-conjugated hybridization probes for preliminary screening of siRNA sequences. *J Am Chem Soc* **127**, 11328–11335.

12. Viel, T., Kuhnast, B., Hinnen, F., Boisgardi, R., Tavitian, B., and Dollé, F. (2007) Fluorine-18 labelling of small interfering RNAs (siRNAs) for PET imaging. *J Labelled Comp Rad* **50**, 1159–1168.

13. Medarova, Z., Evgenov, N. V., Dai, G., Bonner-Weir, S., and Moore, A. (2006) In vivo multimodal imaging of transplanted pancreatic islets. *Nat Protoc* **1**, 429–435.

14. Pham, W., Zhao, B. Q., Lo, E. H., Medarova, Z., Rosen, B., and Moore, A. (2005) Crossing the blood–brain barrier: a potential application of myristoylated polyarginine for in vivo neuroimaging. *Neuroimage* **28**, 287–292.

15. Moore, A., Basilion, J. P., Chiocca, E. A., and Weissleder, R. (1998) Measuring transferrin receptor gene expression by NMR imaging. *Biochim Biophys Acta* **1402**, 239–249.

16. Bartlett, D. W., and Davis, M. E. (2006) Insights into the kinetics of siRNA-mediated gene silencing from live-cell and live-animal bioluminescent imaging. *Nucleic Acids Res* **34**, 322–333.

17. Corish, P., and Tyler-Smith, C. (1999) Attenuation of green fluorescent protein half-life in mammalian cells. *Protein Eng* **12**, 1035–1040.

18. Ward, C. M., and Stern, P. L. (2002) The human cytomegalovirus immediate-early promoter is transcriptionally active in undifferentiated mouse embryonic stem cells. *Stem Cells* **20**, 472–475.

Chapter 2

Targeted Gene Silencing into Solid Tumors with Electrically Mediated siRNA Delivery

Muriel Golzio, Laurent Mazzolini, Aurélie Paganin-Gioanni, and Justin Teissié

Abstract

Short interfering RNAs (siRNAs) represent new potential therapeutic tools, owing to their capacity to induce strong, sequence-specific gene silencing in cells. However, their clinical development requires new, safe, and efficient in vivo siRNA delivery methods. In this study, we report an efficient in vivo approach for targeting gene knockdown in solid tumors by the use of external electric field pulses. We show that gene silencing is efficiently obtained in vivo with chemically synthesized siRNA after targeted electrical delivery in the tumor-bearing mouse. The associated gene silencing was followed on the same animal by fluorescence imaging and confirmed by qPCR. Gene silencing obtained in tumors lasted from 2 to 4 days after a single treatment. Therefore, this method should allow gene function analysis or organ treatment by a localized delivery of siRNAs.

Key words: RNA interference, siRNA, tumor, GFP, fluorescence, in vivo imaging, electroporation, mice, therapy, electropulsation.

1. Introduction

In the last few years, RNAi has become a powerful experimental tool for knocking down the expression of genes of interest *(1)*. Treatment of cells with chemically synthesized small interfering RNAs (siRNAs) is now used as a routine technique for in vitro functional analysis of cellular processes. Furthermore, RNAi-based approaches for clinical applications are currently under intensive development. Among the different limitations encountered in the use of siRNAs for therapy, efficient intracellular

John F. Reidhaar-Olson and Cristina M. Rondinone (eds.), *Therapeutic Applications of RNAi: Methods and Protocols, vol. 555*
© Humana Press, a part of Springer Science+Business Media, LLC 2009
DOI 10.1007/978-1-60327-295-7_2 Springerprotocols.com

uptake and safe in vivo uptake remain critical issues. In addition to different administration modes of the siRNA molecules and chemical modification of the siRNA itself, a variety of chemical and physical methods have been developed to improve the final in vivo delivery of these molecules *(2, 3)*. Among the physical methods, electropulsation (electroporation) has proved to be successful for delivery of siRNAs in a large number of organs and tissues in rodents *(4)*. Electropulsation has indeed been known for more than 25 years to strongly increase in vitro intracellular uptake of molecules and drugs.

In vivo, electric pulses have been extensively used for drug and plasmid delivery in a large number of organs and tissues *(5–7)*. A key feature is that the delivery of molecules is restricted to the volume where the electric field is generated. A physical targeting of the effect is therefore possible. In order to assess the contribution of electrical treatment to siRNA-mediated endogenous gene silencing in solid tumors, we used as a reference experimental model B16-F10 mouse melanoma cell lines stably expressing an enhanced green fluorescent protein (EGFP). The extent of EGFP gene suppression, a model of constitutive gene silencing, in subcutaneous B16-EGFP tumors was subsequently monitored over time in the living animal by whole-body fluorescence imaging.

2. Materials

2.1. Cell Culture

1. B16-F10 mouse melanoma cells and their EGFP-expressing derivatives were obtained after retroviral vector and in vitro transduction (see *(8)* for further details).
2. Cells were cultured in Eagle Minimum Essential Medium (EMEM) (Gibco LifeTechnologies, France) supplemented with 10% fetal calf serum (Gibco), penicillin (100 units/ml), streptomycin (100 mg/ml), and L-glutamine (0.58 mg/ml).
3. Cells were grown on Petri dishes in a 5% CO_2 humidified incubator at 37°C.
4. Dulbecco's PBS buffer (Eurobio, Les Ulies, France) was used to rinse and/or resuspend the cells.
5. Trypsin/EDTA (Eurobio) was used to detach the cells from the Petri dishes.

2.2. siRNAs and DNA Oligonucleotides

1. siRNA Suspension Buffer: 100 mM potassium acetate, 30 mM HEPES-KOH, 2 mM magnesium acetate, pH 7.4.
2. 10 mM Tris-HCl, pH 8.0 (T8.0): autoclaved and stored at 4°C.
3. RNAse inhibitor RNAsin[R] (Promega, Madison, WI).

4. Upon receipt, lyophilized, preannealed double-stranded siR-NAs (Qiagen) are resuspended at a concentration of 100 μM in siRNA Suspension Buffer, heated to 90°C for 1 min, and incubated at 37°C for 60 min. Resolubilized siRNAs are stored at −80°C.

5. The egfp22 siRNA (sense: 5′ r(GCA AGC UGA CCC UGA AGU UCA U), antisense: 5′ r(GAA CUU CAG GGU CAG CUU GCC G)) is directed against GFP mRNA, and was designed according to *(9)* (*see* **Note 1**).

6. The P76 siRNA (sense: 5′ r(GCG GAG UGG CCU GCA GGU A)dTT, antisense: 5′ r(UAC CUG CAG GCC ACU CCG C)dTT) is directed against an unrelated human P76 mRNA and shows no significant homology to mouse transcripts. It is used as a control for specificity of the siRNA construct.

7. Single-stranded DNA oligonucleotides (Sigma) used as primers for quantitative PCR are resuspended in T8.0 buffer at a concentration of 100 μM and stored at −20°C.

8. The primers used for EGFP amplification are EGFP214f (5′-GCA GTG CTT CAG CCG CTA C-3′) and EGFP309r (5′-AAG AAG ATG GTG CGC TCC TG-3′), which were previously described by *(10)*. Primers used for amplification of reference housekeeping genes used for GeNorm calculations are: For *β*-glucuronidase, primers *β*-glucF (5′-ACT TTA TTG GCT GGG TGT GG-3′) and *β*-glucR (5′-AAT GGG CAC TGT TGA TCC TC-3′); for YWHAZ tyrosine 3-monooxygenase, primers mYWHAZ-481F (5′-CGT GGA GGG TCG TCT CAA GT-3′) and mYWHAZ-560R (5′-CTC TCT GTA TTC TCG AGC CAT CTG-3′); for *β* actin-2, primers mActbeta2-5′ (5′-AGC CAT GTA CGT AGC CAT CCA-3′) and mActbeta2-3′ (5′-TCT CCG GAG TCC ATC ACA ATG-3′).

2.3. In Vivo Experiments

1. Female C57Bl/6 mice were purchased from Rene Janvier (St ISLE, France). The C57Bl/6 mice were 9–10 weeks old at the beginning of the experiments, weighing 20–25 g and were considered as young mice. They were maintained at constant room temperature with 12-h light cycle in a conventional animal colony. Before the experiments, the C57Bl/6 mice were subjected to an adaptation period of at least 10 days.

2. Hair removal cream (Veet, Massy, France).

3. Hamilton syringe with a 26G needle.

4. Isoflurane (Forene, Abbott, Rungis, France).

5. Gas anesthesia system composed of an air compressor (TEM, Lormont, France) and an isoflurane vaporizer (Xenogene, Alameda, CA).

6. Electropulsator PS 10 CNRS (Jouan, St Herblain, France). All pulse parameters were monitored on-line with an oscilloscope (Metrix, Annecy, France). An electronic switch cutting the pulse as soon as its intensity reached 5 amp ensured safety against current surge.
7. Plate parallel electrodes (length 1 cm, width 0.6 cm) (IGEA, Carpi, Italy).
8. Conducting paste (Eko-gel, Egna, Italy).

2.4. In Vivo Visualization of Gene Expression and Gene Silencing

1. A fluorescence stereomicroscope (Leica MZFL III, Germany). The fluorescence excitation was obtained with a Mercury Arc lamp (HBO, Osram, Germany) GFP2 filter (Leica).
2. Cooled CCD Camera Coolsnap fx (Roper Scientific, Evry, France).
3. MetaVue software (Universal, USA) drives the camera from a computer and allowed quantitative analysis of the GFP fluorescence level.

2.5. mRNA Extraction and Analysis

1. RNAlater^{TM} RNA stabilization reagent (Qiagen).
2. RNeasy^R RNA isolation kit (Qiagen).
3. FastPrep^R oscillating grinding system and lysing matrix D beads (MP Biomedicals, Solon, OH).
4. RNAse-free water: Add diethylpyrocarbonate (DEPC, Sigma) to deionized water to a final concentration of 0.05% (*see* **Note 2**). Incubate overnight at room temperature and then autoclave for 30 min to eliminate residual DEPC. Store at room temperature.
5. TAE electrophoresis buffer (1X): 40 mM Tris-acetate, pH 8.0, 1 mM Na$_2$EDTA. Prepare as a 50X stock solution, autoclave, and store at room temperature.
6. Ethidium bromide (10 mg/ml solution, Sigma) (*see* **Note 3**).
7. Thermoscript^{TM} RT-PCR system from Invitrogen.
8. β-Mercaptoethanol (molecular biology grade, Sigma).

3. Methods

In this study, electrotransfer of synthetic siRNA against GFP mRNA was used to show the efficiency of in vivo electroadministration after intratumoral injection in solid tumors. We compared treatment groups using egfp22 siRNA, electric field alone, and nonrelevant p76 siRNA. It is important to use the same amount of siRNA among groups as well as the same volumes for injection to obtain reliable results. The first step was to graft the tumor expressing the GFP, and then to determine whether the injection of egfp22 siRNA affects the GFP fluorescence expression. The expression of the fluorescent reporter gene was determined by in vivo fluorescence stereomicroscopy. Since RNAi-mediated gene knockdown acts through RISC-mediated mRNA degradation, validation of gene knockdown was also done by comparing the steady-state levels of target mRNA molecules in control and siRNA-treated samples by quantitative real-time RT-PCR (qPCR).

3.1. Tumor Cells Injection

1. Two days before the injection, an area of at least 1 cm in diameter is shaved on the back of the mouse with the cream (*see* **Note 4**).
2. B16-F10 cells are passaged when confluent with trypsin/EDTA to provide new culture on a 175 cm^2 flask. After 2 days, they are harvested by digestion with trypsin/EDTA and resuspended in PBS buffer at a final dilution of 10^7 cells/ml.
3. A volume of 100 μl of the cell suspension is injected subcutaneously under the shaved area (*see* **Note 5**).
4. The tumor growth is followed daily by measuring the diameter of the tumor by fluorescence imaging (*see* **Note 6**).

3.2. In Vivo Electropulsation

1. Eight to ten days after the subcutaneous injection, the tumor reaches a mean diameter of 5–7 mm (*see* **Note 7**).
2. Anesthetize mice by isoflurane inhalation (*see* **Note 8**).
3. Immediately prior to injection, dilute siRNAs to 17 μM in 50 μl of autoclaved PBS supplemented with 40 units of the RNAse inhibitor RNAsin^R. For all manipulations of the siRNAs, use DNase- and RNase-free buffers and RNase-free plasticware (microtubes and pipette tips). Always wear gloves.
4. Using a Hamilton syringe with a 26-gauge needle, slowly (over about 15 s) inject the siRNA solution into the tumor. In the control "no siRNA" condition, replace the volume of added siRNA solution with the same volume of siRNA Suspension Buffer to keep the injection conditions similar.
5. Apply conducting paste to insure good contact between the skin and the electrodes (*see* **Note 9**).
6. Approximately 30 s after the injection, fit the parallel-plate electrodes around the tumor (*see* **Note 10**) and deliver a train of four pulses plus four additional pulses in the reverse polarity (electrical conditions: 480 V, 20 ms pulse duration, and 1 Hz pulse frequency) (**Fig. 2.1**). Carefully control the delivery of the pulses on the oscilloscope.

3.3. Whole-Body Imaging

Upon electrically mediated transfer, GFP gene expression in the tumor is detected directly on the anesthetized animal by digitized

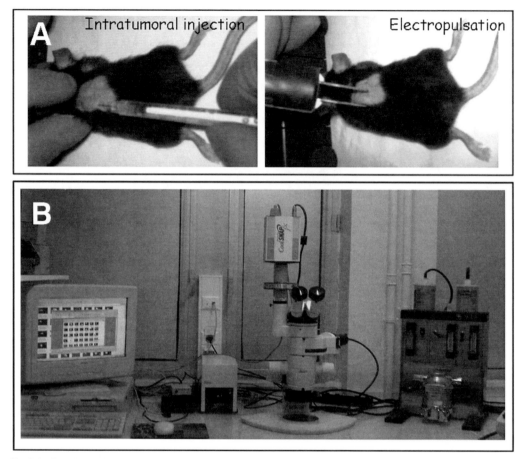

Fig. 2.1. Experimental set-up. **A.** Intratumoral injection of an anesthetized mouse. A volume of 50 μl was injected slowly into the tumor. Electrodes positioned on the tumor allow application of a train of four plus four inverted square wave. **B.** Digitized stereomicroscopy imaging set-up.

stereomicroscopy. The GFP fluorescence from the tumor is quantitatively evaluated at different days (*see* **Fig. 2.2**).

3.3.1. Fluorescence Data Acquisition

1. Anesthetize the mouse.
2. Place the mouse under the stereo-fluorescence microscope, with the tumor on the top of the animal. The whole tumor is observed as a 12-bit, 1.3-M pixel image using a cooled CCD Camera. MetaVue software drives the camera. Take a light picture (*see* **Note 11**).
3. The fluorescence excitation is obtained with a Mercury Arc lamp. The exposure time is set at 1 s with no binning. Acquire by selecting the GFP2 filter (*see* **Note 12**).

3.3.2. Fluorescence Data Analysis

1. The mean fluorescence in the gated area (whole tumor) is quantitatively estimated (measure/region measurement). In our experiments, the background fluorescence is sufficiently

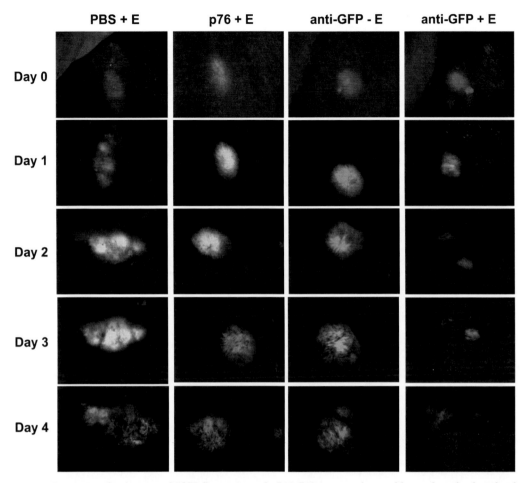

Fig. 2.2. **Representative images of EGFP fluorescence in B16-F10 tumors observed by noninvasive imaging in live animals before (day 0) and after (day 1 to day 4) the different treatments.** Tumors having a mean diameter of 5–7 mm were treated by intratumoral injection of 50-μl saline solution followed by application of electric pulses (PBS + EP); by intratumoral injection of 50 μl of saline solution containing 17 μM of p76 siRNA followed by application of electric pulses (p76 + EP); by intratumoral injection of 50 μl of saline solution containing 17 μM of egfp22 siRNA followed (antiGFP + EP) or not followed (antiGFP – EP) by application of electric pulses. B16-F10 GFP-derived tumors are clearly detected under the animal's skin upon fluorescence excitation and tumor margin can be easily defined. This enables measurement of the tumor area and fluorescence intensity over a period of 4 days after treatment.

low so that it does not interfere with quantitation when GFP emission is present.

2. To quantify the relative time-dependence of knockdown induced by siRNA (**Fig. 2.3**), use the respective intensity of each tumor at day 0 as an animal-specific internal control. The relative fluorescence on day x is represented by the ratio of the fluorescence on day x to this "internal control."

3.4. Statistical Analysis

Six different animals are treated for each condition. Fluorescence level differences between conditions are statistically compared by an unpaired t-test using Prism software (version 4.02, Graphpad).

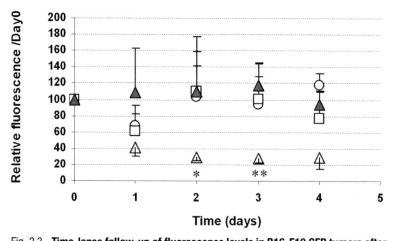

Fig. 2.3. **Time-lapse follow-up of fluorescence levels in B16-F10 GFP tumors after siRNA electrotransfer**. Digital imaging is used to quantify, at different time points, the fluorescence of B16-F10 GFP tumors. On each animal, the mean fluorescence of the tumor is quantitated on a relative scale, using as a reference the fluorescence intensity measured just before treatment (day 0). EGFP fluorescence in tumors following injection of PBS alone with electric field (○); unrelated p76 siRNA electrotransferred (□); egfp22 siRNA without electroporation (▲); and egfp22 siRNA, electrotransferred (△). Vertical bars represent standard deviation. Differences in fluorescence levels between AntiGFP − EP and AntiGFP + EP conditions were statistically compared using an unpaired two-sided *t*-test using KyPlot software. *$p < 0.05$ and **$p < 0.01$ were plotted when observed.

3.5. Extraction of Total RNA from Tumors for RNA Knockdown Validation

3.5.1. Tumor Preparation and RNA Stabilization of Isolated Tumors

1. Process the tumor immediately after recovery. Work as quickly as possible until tumor tissue reaches the RNA stabilization step (*see* **Note 13**).
2. Briefly dip the recovered tumor in ice-cold PBS and dry by rolling the tumor a few seconds on a sterile paper towel.
3. Quickly weigh the tumor. If required, cut the tumor tissue into pieces of up to 50 mg each using a sterile scalpel blade.
4. Place each tumor sample in a 1.5-ml microtube prefilled with at least 10 volumes (i.e., 10 μl/mg of tissue) of RNAlater™ RNA stabilization reagent (making sure that the sample is fully submerged in the solution). Incubate samples overnight at 4°C prior to RNA extraction. For long-term storage, place the tubes at −20°C after an initial overnight incubation at 4°C.

3.5.2. Total RNA Extraction from Stabilized Tissues

1. In order to avoid possible RNA degradation, the duration time of manipulation of the stabilized tumor should be kept to a minimum. Tumors should therefore be processed one

at a time until the homogenization step in Qiagen RNeasy^R RLT lysis buffer has been completed.

2. Prior to extraction, prefill commercial lysing matrix D tubes with 800 μl of RNeasy^R RLT lysis buffer and 8 μl β-mercaptoethanol. Vortex the tubes for a few seconds.

3. Recover the stabilized tumor (stored at 4°C or −20°C) from the RNAlater solution using forceps, and briefly dry it by rolling on a sterile paper towel.

4. Add tumor to the lysing matrix D tube containing RLT lysis buffer and immediately proceed to sample homogenization for 25 s at a setting of 6.0 using the FastPrep^R grinding device.

5. Cool the lysing matrix D tube for 30 s on ice, and then store tube at room temperature.

6. When all tumors have been homogenized, centrifuge lysing matrix D tubes for 3 min at 12,000×g at room temperature.

7. Transfer supernatants in 1.5-ml microtubes by pipetting, and recentrifuge for another 1 min at 12,000×g.

8. Transfer supernatants to new 1.5-ml microtubes, avoiding any turbid material (if present).

9. Process to RNA purification according to Qiagen's RNeasy^R Mini Handbook protocol without any modification (*see* **Note 14**).

3.5.3. Determination of RNA Yield and Control of Integrity

1. For handling of purified RNA, use DNase- and RNase-free buffers, RNase-free plasticware, and wear gloves (*see* **Note 15**).

2. Determine RNA yield by measuring absorbance at 260 mm (OD$_{260}$) on a 100-fold dilution of the sample in RNase-free water using the formula OD$_{260}$ × 4 = μg RNA per μl.

3. RNA integrity is routinely checked by agarose gel electrophoresis under native conditions (*see* **Note 16**). Prepare autoclaved 1X TAE for the gel and electrophoresis buffer. Prepare a 1% agarose gel in autoclaved 1X TAE electrophoresis buffer containing 0.5 μg/ml ethidium bromide.

4. Load 1–2 μg purified RNA in 1X RNase-free loading buffer. Run the gel at up to 10 V/cm until the bromophenol blue dye reaches the lower third of the gel.

5. Photograph under UV illumination (no destaining is required). Under these conditions, two high-intensity, sharp, major bands corresponding to the 28S and 18S ribosomal RNAs should be detected (**Fig. 2.4**). By visual inspection, no significant smearing (indicating partial degradation) should be visible below the rRNA bands, and no single band at the top of the gel (indicating DNA contamination) should be present. If quantitation of RNA bands is possible with the device used for gel image acquisition, the 28S and 18S

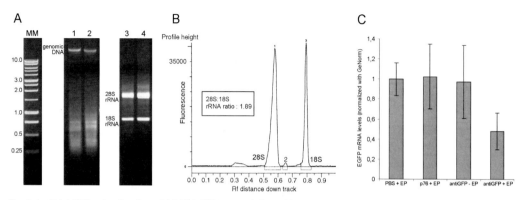

Fig. 2.4. Total RNA extraction from B16-F10 GFP tumors followed by quantitative RT-PCR analysis of EGFP mRNA levels. (**A**) Native gel agarose electrophoresis of total RNA extracted from tumors. Lanes 1 and 2 show examples of RNA initially obtained from tumors by using poorly denaturing RNA extraction buffers and moderately efficient grinding conditions. Degraded RNA appears as a smear in the gel. Note the presence of contaminating genomic DNA at the top of the gel. Lanes 3 and 4 show total RNA extracted from tumors using the described protocol. Two major bands corresponding to 28S and 18S ribosomal RNAs are present. No contaminating genomic DNA is detectable. MM: molecular mass marker. (**B**) Quantification of ethidium bromide fluorescence of sample shown in lane 3. 28S and 18S rRNAs give sharps peaks with a 28S:18S ratio close to 2.0 indicative of intact RNA. (**C**) EGFP reporter gene steady-state transcript levels were determined by qPCR analysis of RNA extracted from tumors 4 days after treatments with: electric field (PBS + EP), unrelated siRNA, electrotransferred (p76 + EP), egfp22 siRNA without electrotransfer (AntiGFP − EP), and egfp22 siRNA followed by electrical treatment (AntiGFP + EP). EGFP transcript levels were determined using a normalization factor calculated from three independent constitutive housekeeping genes using the GeNorm VBA applet. Three independent tumors were analyzed in each case. Histograms represent mean value ± standard deviation.

rRNAs should appear as sharp peaks with 28S:18S quantitative ratios of 1.8–2.0.

3.5.4. qPCR Analysis of Target mRNA Levels

1. Total RNA (5 μg) isolated from the different tumor samples is used to generate cDNAs through reverse transcription by the Thermoscript™ reverse transcriptase (Invitrogen) using random hexamers as primers and subsequent RNase H treatment.

2. qPCR is then performed on 5 μl of a 25-fold dilution of the cDNA synthesis reaction using an ABI Prism 7000 system (Applied Biosystems) and SYBR Green dye fluorescence measurements to quantify amplicon production.

3. Three independent endogenous mouse genes giving the most highly reproducible constitutive expression levels in tumors (*see* **Section 2.2**) are selected over a panel of tested "housekeeping" mouse genes using the *GeNorm* Visual Basic application for Microsoft Excel *(11)*. This application is freely accessible on the Web for noncommercial academic research.

4. The normalization factor calculated by GeNorm from the selected genes is then used to determine target gene mRNA levels.

4. Notes

1. Selection of the optimal siRNA sequences remains a critical issue for RNAi experiments in vivo (for a general discussion *see (12, 13)*). Literature mining may be a valuable source for the identification of "validated" active siRNAs directed against a gene of interest. In parallel, most companies commercializing siRNAs now propose various sets of "predesigned siRNAs" against any mouse or human gene, which are designed by robust, proprietary, algorithms. Similar freely accessible algorithms can also be found on the Internet (e.g., *(13–15)*). A validation and selection of the most active siRNAs in cultured cells should always precede their in vivo use.

2. DEPC is highly toxic and volatile. It must be used only in a laboratory chemical fume hood.

3. Ethidium bromide is a potent mutagen. Always wear gloves and minimize handling. Use specific procedures (such as charcoal filtration) for disposal of ethidium bromide-containing buffers. Use dedicated electrophoresis tanks.

4. The cream should be used 2 days before fluorescence imaging because some components of the cream fluoresce under blue-light excitation. This cream should be used carefully as it may cause some irritations in the mouse skin. Rinse the cream under a flow of water.

5. The subcutaneous injection should be performed under the shaved area to allow direct visualization of the GFP-expressing cells under the skin.

6. GFP expression in the tumor cells is detected directly through the skin on the anesthetized animal by digitized fluorescence stereomicroscopy. This procedure allows observation of GFP expression in the same animal over several days.

7. Hair on the back may re-grow; re-shave the skin above the tumor if necessary.

8. Isoflurane inhalation is safe; mice recover very quickly after the electrical treatment and can be observed daily for in vivo imaging.

9. Conducting paste is very important to insure optimal electrical contact with the skin. One should pay attention that the paste is not continuous between the two electrodes, as the field will pass through the paste and not through the tumor.

10. One person should perform the tumor injection. As the tumor may have different nodules, pay attention when injecting the whole tumor.

11. One person should be responsible for holding the mice to avoid erratic conditions of exposure to the fluorescence excitation beam.

12. Make sure that no saturation of pixels occurs in the area of interest upon image acquisition. Then use identical settings for subsequent acquisitions.

13. In our experiments, extraction of nondegraded RNA from tumors required use of tumor tissue pretreated with an RNA stabilizing agent together with highly denaturing RNA extraction buffers (containing guanidine isothiocyanate), as well as harsh homogenization conditions. We assume that this may be due to the presence of necrotic regions in the tumors, which may release high levels of nucleases in the tissue. Although the reagents and materials used in this protocol may be substituted with others, we recommend using extraction conditions that fulfill the criteria mentioned above.

14. Contamination of the purified RNA sample by genomic DNA sequences may be detrimental to sensitive applications such as qPCR. However, in our hands, additional treatments performed in order to eliminate putative DNA contamination did not modify the qPCR amplification patterns.

15. The electrophoresis tank should have been pretreated with an RNase-removal reagent (e.g., RNaseZap^R RNase decontamination solution from Ambion) to avoid any RNA degradation during electrophoresis due to contaminating RNases.

16. Although denaturing conditions provide the greatest resolution for RNA analysis, direct electrophoresis under native conditions was found sufficient to assess the integrity and overall quality of purified RNA. In addition, native RNA staining is much more sensitive than that of denatured DNA.

Acknowledgments

The authors would like to thank laboratory members for their comments, Bettina Couderc for providing us the GFP-expressing B16 cells, and Marie-Jeanne Pilaire for helpful discussion on qPCR. Financial support was obtained from the Ligue Nationale Contre le Cancer, the CNRS CEA Imagerie du Petit Animal Program, the Région Midi-Pyrénées, the Cancéropôle GSO (Grand Sud-Ouest), and the AFM (Association Française contre les Myopathies).

References

1. Rana, T.M. (2007) Illuminating the silence: understanding the structure and function of small RNAs. *Nat Rev Mol Cell Biol.* **8**, 23–36.
2. de Fougerolles, A., et al. (2007) Interfering with disease: a progress report on siRNA-based therapeutics. *Nat Rev Drug Discov.* **6**, 443–453.
3. Kim, D.H. and J.J. Rossi. (2007) Strategies for silencing human disease using RNA interference. *Nat Rev Genet.* **8**, 173–184.
4. Behlke, M.A. (2006) Progress towards in vivo use of siRNAs. *Mol Ther.* **13**, 644–670.
5. Heller, L.C., K. Ugen, and R. Heller (2005) Electroporation for targeted gene transfer. *Expert Opin Drug Deliv.* **2**, 255–268.
6. Li, S. (2004) Electroporation gene therapy: new developments in vivo and in vitro. *Curr Gene Ther.* **4**, 309–316.
7. Wells, D.J. (2004) Gene therapy progress and prospects: electroporation and other physical methods. *Gene Ther.* **11**, 1363–1369.
8. Golzio, M., et al. (2007) In vivo gene silencing in solid tumors by targeted electrically mediated siRNA delivery. *Gene Ther.* **14**, 752–759.
9. Caplen, N.J., et al. (2001) Specific inhibition of gene expression by small double-stranded RNAs in invertebrate and vertebrate systems. *Proc Natl Acad Sci USA.* **98**, 9742–9747.
10. Klein, D., et al. (2000) Accurate estimation of transduction efficiency necessitates a multiplex real-time PCR. *Gene Ther.* **7**, 458–463.
11. Vandesompele, J., et al. (2002) Accurate normalization of real-time quantitative RT-PCR data by geometric averaging of multiple internal control genes. *Genome Biol.* **3**, 34.
12. Birmingham, A., et al. (2007) A protocol for designing siRNAs with high functionality and specificity. *Nat Protoc.* **2**, 2068–2078.
13. Pei, Y. and T. Tuschl (2006) On the art of identifying effective and specific siRNAs. *Nat Methods.* **3**, 670–676.
14. Vert, J.P., et al. (2006) An accurate and interpretable model for siRNA efficacy prediction. *BMC Bioinformatics.* **7**, 520.
15. Gong, W., et al. (2006) Integrated siRNA design based on surveying of features associated with high RNAi effectiveness. *BMC Bioinformatics.* **7**, 516.

Chapter 3

Liposomal siRNA for Ovarian Cancer

Lingegowda S. Mangala, Hee Dong Han, Gabriel Lopez-Berestein, and Anil K. Sood

Abstract

Discovery of RNA interference (RNAi) has been one of the most important findings in the last ten years. In recent years, small interfering RNA (siRNA)-mediated gene silencing is beginning to show substantial promise as a new treatment modality in preclinical studies because of its robust gene selective silencing. However, until recently, delivery of siRNA in vivo was a major impediment to its use as a therapeutic modality. We have used a neutral liposome, 1,2-dioleoyl-*sn*-glycero-3-phosphatidylcholine (DOPC), for highly efficient in vivo siRNA delivery. Using siRNA tagged with Alexa-555, incorporated in DOPC liposomes, we have demonstrated efficient intra-tumoral delivery following either intraperitoneal or intravenous injection. Furthermore, EphA2-targeted siRNA in DOPC liposomes showed significant target modulation and anti-tumor efficacy.

Key words: RNA interference, siRNA delivery, neutral liposome, DOPC, ovarian carcinoma, EphA2.

1. Introduction

RNA interference (RNAi) is thought to have evolved as a form of innate immunity against viruses and has become a powerful tool for highly specific gene silencing and drug development. RNAi is a mechanism for RNA-guided regulation of gene expression in which double-stranded ribonucleic acid (dsRNA) inhibits the expression of genes with complementary nucleotide sequences. Long dsRNA is cleaved by Dicer, which is an RNAse III family ribonuclease. This process yields small interfering RNAs (siRNAs) about 21 nucleotides long. These siRNAs are incorporated into a multi-protein RNA-induced silencing complex (RISC) that

John F. Reidhaar-Olson and Cristina M. Rondinone (eds.), *Therapeutic Applications of RNAi: Methods and Protocols, vol. 555*
© Humana Press, a part of Springer Science+Business Media, LLC 2009
DOI 10.1007/978-1-60327-295-7_3 Springerprotocols.com

is guided to target mRNA, which is then cleaved *(1–4)*. In mammalian cells, the related microRNAs (miRNAs) are found, which are short, ~22 nucleotides long, RNA fragments. These miRNAs are generated after Dicer-mediated cleavage of longer (~70 nucleotide) precursors with imperfect hairpin RNA structures. The miRNA is incorporated into a miRNA–protein complex (miRNP), which leads to translational repression of target mRNA.

siRNA-mediated gene silencing offers options for targeting genes that cannot be targeted with other approaches such as small-molecule inhibitors or monoclonal antibodies. However, for clinical development, two major barriers must be overcome, nuclease digestion and in vivo delivery. Several approaches including plasmids and viral vectors have been used *(5)* with some successful in vitro applications, but limited potential for in vivo use *(6)*. Liposomes are self-assembled, closed structures composed of lipid layers formed spontaneously upon the addition of water to a dried lipid film. We have shown that nano-liposomes can serve as an effective vehicle for the delivery of siRNA. Liposomal formulations have been used to incorporate and deliver a wide variety of therapeutic and diagnostic agents. The circulating half-life of liposomes can be prolonged by the addition of neutral, hydrophilic polymers such as poly(ethylene glycol) (PEG) to the outer surface *(7)*. An extended circulation half-life allows for sustained availability in order to take advantage of the enhanced permeability of tumor vasculature, resulting in increased delivery to target sites *(8–11)*. Moreover, it offers opportunities to further enhance the specificity and efficiency of siRNA delivery to tumor tissue, by appropriately "flagging" the liposomal surface with molecular tags, to minimize the non-specific toxic effects.

Our goal was the development of a safe, biodegradable, and biocompatible carrier system for siRNA that will protect siRNA from nuclease digestion and increase delivery into tumors. Liposomes, in general, have been shown to be safe in a number of clinical trials using a wide variety of anticancer and antimicrobial drugs. The liposomes' charge, size, and lipid composition will impact their safety, distribution, and uptake by cells and tissues. For example, negatively charged liposomes will be avidly taken up by phagocytic cells and may not result in optimal loading efficiency due to the negative charge of the siRNA *(12)*. Cationic liposomes have low delivery efficiency because of their electrostatic interactions with other cells and biomolecules along the delivery path. Therefore, we have focused on the use of a neutral liposome, 1,2-dioleoyl-*sn*-glycero-3-phosphatidylcholine (DOPC), for in vivo delivery of siRNA.

We have recently demonstrated that siRNA incorporated into nano-liposomes, composed of DOPC, are able to penetrate deep into tumors in vivo while avoiding phagocytic uptake (we will use the term DOPC to mean liposomes composed of DOPC). For these initial delivery studies, we used Alexa-555-tagged siRNA. Our initial proof-of-concept studies focused on the use of DOPC for targeting a frequently overexpressed tyrosine kinase, EphA2, in ovarian cancer *(13–15)*. EphA2 is a tyrosine kinase receptor belonging to the ephrin family that plays a key role in neuronal development. In adults, it is expressed to a low degree, primarily in epithelial cells *(14)*. This differential expression in tumor cells makes EphA2 an attractive therapeutic target. We demonstrated that EphA2 siRNA-DOPC given either i.v. or i.p. resulted in EphA2 silencing in orthotopic experimental ovarian tumors, leading to reduced tumor growth as a single agent or in combination with a taxane *(14, 16)*.

2. Materials

2.1. Cell Culture

1. Roswell Park Memorial Institute-1640 (RPMI) Medium supplemented with 15% fetal bovine serum (FBS) and 0.1% gentamycin.
2. Trypsin solution (0.25%) and ethylenediamine tetraacetic acid (EDTA, 1 mM).
3. Hank's Balanced Salt Solution (Serum free) with calcium and magnesium.
4. BCA assay kit.

2.2. Transfection of Cancer Cells with siRNA

1. A non-silencing fluorescent siRNA sequence tagged with Alexa-555, 5′-AATTCTCCGA-ACGTGTCACGT-3′, control siRNA (same sequence without Alexa-555), and EphA2-targeted sequence 5′-AATGACATGCCGATCTACATG-3′ were synthesized and purified.
2. siRNA suspension buffer (100 mM potassium acetate, 30 mM HEPES-potassium hydroxide, 2 mM magnesium acetate, pH 7.4).
3. RNAiFect transfection reagent.
4. 1, 2-Dioleoyl-*sn*-glycero-3-phosphatidylcholine (DOPC).
5. Molecular weight limit filters.

2.3. Western Blotting

1. Modified radioimmunoprecipitation assay (RIPA) lysis buffer: 50 mM Tris, 150 mM NaCl, 1% Triton, 0.5% deoxycholate plus protease inhibitors (25 μg/mL leupeptin, 10 μg/mL aprotinin, 2 mM EDTA, 1 mM sodium orthovanadate).
2. Running buffer: Prepare 10X stock with 1.92 M glycine, 250 mM Tris, 1% SDS. Dilute 100 mL with 900 mL distilled water (1X) for use.

3. Transfer buffer: Prepare 10X stock of glycine and Tris as said above. Add 20% methanol plus 0.05% (w/v) SDS to 100 mL of stock and make up to 1 L with distilled water.

4. Tris-buffered saline with Tween-20 (TBST): Prepare 10X stock with 100 mM NaCl, 50 mM Tris-base. Adjust pH to 7.5 with concentrated HCl and make up to 1 L with distilled water. Add 0.1% Tween-20 to 100 mL stock and make up to 1 L (1X) before use.

5. Blocking buffer: 5% (w/v) nonfat dry milk in TBST.

6. Stacking gel: 0.5 M Tris-HCl buffer with pH 6.8.

7. Resolving gel: 1.5 M Tris-HCl buffer with pH 8.8.

8. Ready gel blotting sandwich containing 0.45 μm nitrocellulose membrane with filter paper (7 × 8.5 cm).

9. Primary anti-EphA2 antibody in 1% nonfat dry milk in TBST.

10. Anti-beta-actin antibody.

11. Secondary horseradish peroxidase (HRP)-conjugated anti-mouse IgG in 1% nonfat dry milk in TBST.

12. Enhanced chemiluminescent (ECL) detection reagents.

2.4. Immunohisto-chemistry

2.4.1. Determination of Uptake of Alexa-555 Fluorescent siRNA by Tumor Tissues

1. Normal horse and goat serum.

2. Hoechst 33342 trihydrochloride, trihydrate.

3. Rat anti-mouse CD31 primary antibody.

4. Rat anti-mouse f4/80 primary antibody.

5. Goat anti-rat Alexa-488 secondary antibody.

6. Fluorescent mounting medium: Glycerol-propyl galate in PBS.

7. Fluorescent Zeiss Axioplan 2 microscope.

2.4.2. Immunohistochemical Staining for EphA2 in Tumors

1. Mouse IgG Fragment (Fc) blocker.

2. Primary EphA2 antibody (Anti-EphA2 clone EA5, a kind gift from Dr. Michael Kinch).

3. Biotinylated horse anti-mouse antibody.

4. Streptavidin-HRP.

5. 3, 3-Diaminobenzidine substrate.

6. Gill No. 3 hematoxylin.

3. Methods

3.1. Cell Culture

1. Ovarian cancer HeyA8 cells were maintained in log phase of cell growth by culturing in RPMI-1640 medium supplemented with 15% FBS and 0.1% gentamicin sulfate at 37°C in 5% CO_2/95% air.

2. All in vitro experiments were carried out in triplicates using 70–80% confluent cells.

3.2. Preparation of Cells for In Vitro Transient siRNA Transfection

1. The purified siRNAs were dissolved in suspension buffer to a final concentration of 20 μM, heated to 90°C for 60 s, and incubated at 37°C for 60 min prior to use to disrupt any higher aggregates formed during synthesis.
2. To check the downregulation of targeted protein EphA2 in vitro using a target siRNA sequence, in vitro transient transfection was carried out using HeyA8 cells.
3. HeyA8 cells (70–75% confluent) were washed with PBS twice and detached using trypsin-EDTA solution. Trypsin activity was neutralized by RPMI-1640 serum containing media. Cells were counted using trypan blue and 3.5×10^5 cells were seeded on six-well tissue culture dishes. The cells should be plated 24 h prior to transfection and avoid stressing the cells without medium during washing steps (*see* **Note 1**). After 24 h, transfection was carried out using RNAiFect transfection reagent. SiRNA (5 μg) was mixed with 65 μL RPMI-1640 media with serum and 30 μL RNAiFect transfection reagent (*see* **Note 2**). SiRNA and transfection reagent mixture was incubated for 15–20 min to allow for the formation of RNA–lipid complex at room temperature and then mixture was added on to cells. Cells were collected after 48 h to check the downregulation of EphA2 using Western blotting.

3.3. Preparation of Liposomal siRNA for In Vivo Delivery

1. For in vivo delivery studies, siRNA was incorporated into DOPC.
2. siRNA and DOPC were mixed in the presence of excess tertiary butanol at a ratio of 1:10 (w/w) siRNA/DOPC. The mixture along with Tween-20 was vortexed, frozen in an acetone/dry ice bath, and lyophilized.
3. To estimate the amount of siRNA not taken up by liposomes (unbound siRNA), free siRNA was separated from liposomes by centrifuging the liposomal-siRNA suspension using 30,000 nominal molecular weight limit filters at $5,000 \times g$ for 40 min at room temperature. Fractions were collected and the material trapped in the filter was reconstituted with 0.9% NaCl. siRNA in the collected fraction and elute was measured by spectrophotometer.
4. Before in vivo administration, to achieve the desired dose in 200 μL per injection per mouse, lyophilized liposomal siRNA complex was suspended in 0.9% NaCl.

3.4. Orthotopic In Vivo Model of Advanced Ovarian Cancer

1. Female athymic nude mice (NCr-nu) were purchased from the National Cancer Institute-Frederick Cancer Research and Development Center (Frederick, MD).

2. Animals were housed in specific pathogen-free conditions and taken care according to the guidelines of the American Association for Accreditation of Laboratory Animal Care and the USPHS "Policy on Human Care and Use of Laboratory Animals".

3. All studies were approved and supervised by the University of Texas M.D. Anderson Cancer Center Institutional Animal Care and Use Committee.

3.5. Preparation of Cells for In Vivo Injections

Tumors were established by i.p. injection of cells prepared as below. For in vivo injection, 60–80% confluent HeyA8 cells were trypsinized and centrifuged at 1,000 rpm for 7 min at 4°C (*see* **Note 3**). Then cells were washed twice with PBS, and reconstituted in serum-free HBSS at a concentration of 1.25×10^5 cells/mL to inject 200 μL i.p. injections per animal. Tumors established in the animals by using this method reflect the i.p. growth pattern of advanced ovarian cancer *(17, 18)*. For treatments, siRNA–DOPC complex was mixed gently with PBS or 0.9% NaCl with pipetting up and down before injection (*see* **Note 4**). Do not freeze the liposomal-siRNA complex after suspending with PBS or normal saline (*see* **Note 5**).

3.6. Determination of Uptake and Distribution of Fluorescent siRNA in Ovarian Tumors

1. A non-silencing fluorescent siRNA sequence tagged with Alexa-555 (5′-AATTCTCCGAACGTGTCACGT-3′), which did not share any sequence homology with any known human mRNA sequences, was used to determine uptake and distribution in various tissues when given in vivo. Same sequence without an Alexa-555 tag was used as control siRNA for tissue background.

2. Determination of uptake of a single dose of fluorescent siRNA in tumor and different organs was initiated once i.p. tumors reached a size of 0.5–1.0 cm^3 as assessed by palpation (~17 days after tumor cells injection).

3. Non-silencing Alexa-555 siRNA–DOPC or control siRNA–DOPC (5 μg) in 100 μL of normal saline was given through i.v. bolus into the tail vein under normal pressure.

4. Tumor and other tissues were harvested at various time points after injection (1 h, 6 h, 48 h, 4 days, 7 days, or 10 days).

5. Tissue specimens were placed in OCT and frozen rapidly in liquid nitrogen for frozen slide preparation.

3.7. Determination of EphA2 Downregulation in Ovarian Tumors

1. To characterize the efficacy of targeted delivery of siRNA packaged into neutral nano-liposomes, HeyA8 ovarian cancer cells were used. A target siRNA sequence, 5′-ATGACATGCCGATCTACATG-3′, was used to downregulate EphA2 in vitro and in vivo.

2. To determine the optimal dose and frequency of dosing, a single injection of EphA2 siRNA–DOPC was given i.v. once tumor size reached 0.5–1.0 cm³, as assessed by palpation (~17 days after tumor cells injection).

3. Targeted EphA2 siRNA–DOPC or control siRNA–DOPC (5 μg in 100 μL normal saline) was given through i.v. bolus into the tail vein under normal pressure.

4. Tumors were harvested at several time points following siRNA injection. Tissue specimens were either snap frozen for protein assessment, fixed in formalin for paraffin embedding, or frozen in OCT for immunofluorescence.

5. To assess tumor growth for long-term experiments, therapy was started one week after tumor cell injection. Mice were divided into five groups ($n = 10$ per group): (a) empty liposomes; (b) non-targeting control siRNA-DOPC; (c) EphA2 siRNA-DOPC; (d) control siRNA-DOPC + paclitaxel; and (e) EphA2 siRNA-DOPC + paclitaxel. Paclitaxel (100 μg/mouse) was injected i.p. once weekly and siRNAs (non-specific or EphA2-targeted, 150 μg/kg) injected i.v. twice weekly.

6. Mice were monitored for adverse effects. When mice in any group began to appear moribund (approximately after 3–4 weeks of therapy), all animals were sacrificed and tumors were harvested.

7. Mouse weight, tumor weight, and distribution of tumor were recorded. Tumor tissue specimens were collected for Western blotting and immunohistochemistry studies as described above. Results are shown in **Fig. 3.1**.

3.8. Western Blotting

1. HeyA8 cells were grown to 80% confluence, washed two times with PBS, and then lysed in modified RIPA buffer plus protease inhibitors on ice for 20 min.

2. Cell lysates were scraped off, kept on ice for 20 min, and centrifuged at 12,000 rpm for 20 min at 4°C.

3. The protein content of the supernatants was measured by the BCA assay kit.

4. The cell lysates were adjusted for equal amount of protein, mixed with 3X sample buffer, and boiled for 5 min.

5. Thirty micrograms of cell lysate protein was separated by sodium dodecyl sulfate-polyacrylamide gel electrophoresis (10% gel) and electrophoretically transferred on to a nitrocellulose membrane.

6. The membrane was blocked overnight in 5% (w/v) nonfat dry milk in TBS containing 0.1% Tween-20.

7. The membrane was probed with an EphA2 antibody at 1:1,000 dilution overnight at 4°C and washed with TBST for 1 h at room temperature by changing the TBST every 10 min.

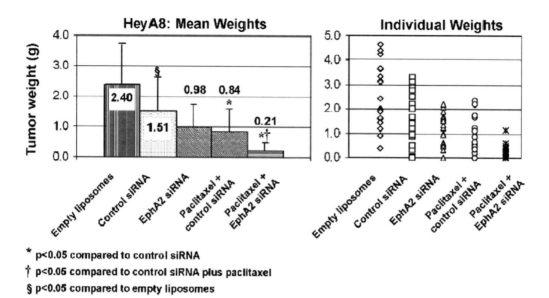

* p<0.05 compared to control siRNA

† p<0.05 compared to control siRNA plus paclitaxel

§ p<0.05 compared to empty liposomes

Fig. 3.1. Therapeutic effects of siRNA-mediated EphA2 downregulation on HeyA8 ovarian tumors. A total of 2.5 × 10⁵ HeyA8 cells were injected i.p. into nude mice. Mice were randomly divided into five groups: (**A**) empty liposomes, (**B**) control siRNA–DOPC, (**C**) EphA2 siRNA–DOPC, (**D**) control siRNA–DOPC + paclitaxel, and (**E**) EphA2 siRNA–DOPC + paclitaxel. Therapy was started on the seventh day by injecting liposomal siRNA twice weekly (150 μg/kg) and paclitaxel (100 μg/mouse) once weekly. Animals were killed when control mice became moribund (3–5 weeks after cell injection); mouse weight, tumor weight, and tumor location were recorded. On left, mean tumor weight, and on right, individual tumor values were shown. (Reproduced from Ref. *(16)* with permission from American Association for Cancer Research).

8. The membrane was reprobed with secondary anti-mouse IgG HRP at 1:2,000 dilution.

9. Secondary antibody was removed and membrane was washed for 1 h at room temperature by changing the TBST every 10 min.

10. The enhanced chemiluminescence system was used to detect the reaction between antigen and antibody. After final wash with PBS, remaining steps were done in a dark room. Aliquots (1 mL) of each portion of the ECL reagent was mixed and added onto the blot and then rotated by hand for 1 min to ensure even coverage of the ECL reagent. The blot was removed from the ECL reagent and blotted.

11. The membrane was placed in X-ray film cassette with film for a suitable exposure time, typically a few minutes.

12. The membrane was stripped and reprobed with an anti-beta-actin antibody at a dilution of 1:2,000 to ensure even loading of proteins in the different lanes (**Fig. 3.2A**).

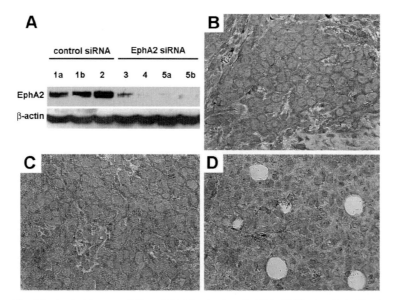

Fig. 3.2. In vivo downregulation of EphA2 using target siRNA. (**A**) Western blot of lysate from tumors collected after 48 h of single administration of control siRNA–DOPC (lanes 1 and 2) or EphA2-targeting siRNA–DOPC (lanes 3–5). Lanes 1a and 1b are separate preparations from same tumor treated with control siRNA–DOPC. Similarly 5a and 5b are from same tumor treated with EphA2-targeted siRNA–DOPC with different preparations. Lanes 2–4 are from separate mice treated with either control or EphA2 siRNA with DOPC. (**B**) Immunohistochemical staining for EphA2 in tumor treated with control siRNA–DOPC. (**C**) Immunohistochemistry 48 h after a single injection of EphA2-targeted siRNA without transfection agent (naked siRNA) is shown and had no detectable effect on EphA2 expression. (**D**) Treatment with EphA2-targeted siRNA in DOPC showed effective downregulation of EphA2 48 h after single injection. B–D, original magnification ×400. (Reproduced from Ref. *(16)* with permission from American Association for Cancer Research).

3.9. Immunohisto-chemistry

3.9.1. Immunohistochemical Staining for EphA2

1. Sections of formalin-fixed, paraffin-embedded tumor samples (8-μm thick) were heated to 60°C.
2. Sections were dehydrated in xylene, 100% ethanol, 95% ethanol, and 80% ethanol.
3. Then, tissues were rehydrated with PBS.
4. Antigen retrieval was done with 0.2 mol/L Tris/HCl (at pH 9.0) for 20 min in a steam cooker.
5. Slides were allowed to cool for 20 min at room temperature, followed by repeated rinsing with PBS for three times for 5 min.
6. Individual slides were removed, and dried around the tissue. A circle was drawn around the tissue with a Pap pen and kept on the slide holder. A drop of PBS was added to tissue and

placed in a humidified chamber. This step was performed for all immunohistochemistry and confocal microscopic studies.

7. Endogenous peroxidase activity was quenched with 3% hydrogen peroxide in methanol for 5 min and washed three times with PBS for 5 min.

8. Slides were counterstained with hematoxylin and mounted. The negative controls were incubated with non-immune mouse IgG in place of primary antibody (**Fig. 3.2B–D**).

3.9.2. Determination of Uptake of Alexa-555 Fluorescent siRNA by Tumor Tissues

1. Tissues were collected from sacrificed mice, immediately placed in OCT, and frozen rapidly in liquid nitrogen (**Section 3.6**).

2. Frozen sections were cut at 8-μm thickness and fixed with fresh cold acetone for 10 min.

3. Tissues were washed three times with PBS for 5 min and slides were placed on the slide holder as mentioned above in **Section 3.9A**. The humidified chamber was covered with an aluminum foil to protect from light.

4. Samples were exposed to 1.0 μg/mL Hoechst in PBS (at 1:10,000 dilution) for 10 min to stain the nucleus.

5. After three washes for 5 min with PBS, tissues were covered with mounting medium glycerol-propyl-gallate in PBS, coverslipped, and examined with a fluorescent Zeiss Axioplan 2 microscope, Hamamatsu ORCA-ER Digital camera, and ImagePRO Software using red and blue filters for Alexa-555 siRNA and nuclei, respectively (**Fig. 3.3A**). Since the fluorescent signal becomes weaker over time, pictures were taken within 4 days (*see* **Note 6**).

3.9.3. Immunofluorescence Detection of Macrophages

1. OCT frozen sections of 8 μm thickness were fixed with fresh cold acetone for 10 min and washed three times with PBS for 5 min.

2. Slides were placed in a humidified chamber as mentioned above and blocked with protein block for 20 min at room temperature.

3. Extra protein block was drained off and tissues were incubated with 10 μg/mL anti-f4/80 primary macrophage antibody at 4°C overnight.

4. Tissues were washed with PBS three times for 5 min each and then re-incubated with 4 μg/mL secondary goat anti-rat Alexa-488 for 1 h at room temperature.

5. Tissues were washed three times with PBS and counterstained with Hoechst, which was diluted 1:10,000 in PBS.

6. Tissues were washed with PBS for three times, mounted using mounting media glycerol-propyl-gallate in PBS, and then coverslipped.

7. Samples were visualized using conventional fluorescence microscope using appropriate filters such as red filter for

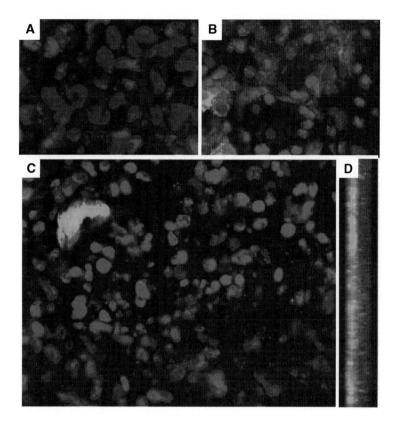

Fig. 3.3. In vivo siRNA distribution in HeyA8 tumor (i.p.) after a single siRNA dose. Tumors were harvested at different time points, frozen in OCT medium, fixed in acetone, and exposed to Hoechst to stain nuclei *blue*. (**A**) Fluorescent siRNA distribution in tumor tissue (original magnification ×400). (**B** and **C**) To detect scavenger macrophages and endothelial cells, tumor tissues were processed as mentioned above, additionally exposed to anti-f4/80 and anti-CD31 antibody and then to Alexa-488-tagged secondary antibody (*green*). (**D**) Sections of 30 μm thickness were examined with confocal microscopy. Photographs taken every 1 μm were stacked and examined from the lateral view. Nuclei were labeled blue and fluorescent siRNA (*red*) was seen throughout the section. Fluorescent Alexa-488-tagged secondary antibody (*green*) is trapped on the surface since it is too large to penetrate the tissue. (Reproduced from Ref. *(16)* with permission from American Association for Cancer Research).

fluorescent siRNA, green for macrophage, and blue for nucleus (**Fig. 3.3B**).

3.9.4. Evaluation of the Tumor Vasculature

1. OCT frozen sections were cut at 8 μm thickness.
2. Tissues were fixed with fresh cold acetone for 10 min and then washed with PBS three times for 5 min.
3. Tissues were placed in humidity chamber on the slide holder as mentioned above in **Section 3.9A**.
4. Tissues were incubated with protein block (5% normal horse serum plus 1% normal goat serum in PBS) for 20 min.

5. Excess of protein block was removed and tissues were incubated with primary rat anti-mouse CD31 antibody, which was diluted in protein block at dilution of 1:800 at 4°C overnight.
6. Slides were washed three times with PBS for 5 min each and incubated with secondary goat anti-rat Alexa-488 (1:400 dilution) for 1 h at room temperature. The humidity chamber was covered with aluminum foil to protect from light.
7. Tissues were washed three times with PBS for 5 min and counterstained with Hoechst (1:10,000 dilution) for 10 min for the detection of nucleus.
8. Tissues were mounted using propyl-gallate and coverslipped.
9. Samples were visualized for CD31-positive blood vessels by fluorescent Zeiss Axioplan 2 microscope, Hamamatsu ORCA-ER Digital camera, and ImagePRO Software using red, blue, and green filters for Alexa-555 siRNA, nuclei, and blood vessels, respectively (**Fig. 3.3C**).

3.10. Confocal Microscopy

1. For confocal microscopy, tissues were collected from sacrificed mice, immediately placed in OCT, and frozen rapidly in liquid nitrogen.
2. Frozen sections were cut at 30-μm thickness, fixed with fresh cold acetone for 10 min, and washed three times with PBS for 5 min.
3. Sections were exposed with 10 nmol/L Sytox green in PBS for 10 min to stain the nucleus.
4. After three washes for 5 min with PBS, tissues were covered with glycerol-propyl-gallate, coverslipped, and examined with Zeiss LSM 510 confocal microscope and LSM 510 Image Browser software (**Fig. 3.3D**).

4. Notes

1. Avoid stressing cells with temperature shifts and long periods without medium during washing steps. It is very important that the cell density is not too low at transfection. Ideally, the cells should be plated 24 h prior to transfection.
2. For in vitro siRNA transfection of ovarian cancer cells, we optimized the ratio between siRNA and RNAiFect as 1:6. If there is toxicity seen for other cell types with this ratio, we recommend decreasing the amount of RNAiFect reagent.
3. For in vivo injections, grow cells to 60–80% confluence. Do not use over-confluent cells and cells that have propagated more than 10 passages (since the last thaw), as they will have lower take rates.

4. Mix siRNA–DOPC with PBS (without calcium or magnesium) or normal saline before injection of the complex into mice. Do not vortex the siRNA–liposome complex (degradation of liposomes may occur with vortexing). Mix gently with pipetting up and down.

5. Once liposomal siRNA is mixed with PBS or saline, do not freeze the solution because the freezing process could fracture or rupture the liposomes leading to a change in size distribution and loss of internal contents.

6. When Alexa-555 fluorescently labeled siRNA is used, we recommend taking pictures within 3–4 days after staining because the fluorescent signal may become weaker over time.

Acknowledgments

Portions of the work in this paper were supported in part by a grant from the Department of Defense (W81XWH-04-1-0227), NIH grants (CA109298, CA110793), the U.T.M.D. Anderson Cancer Center SPORE in ovarian cancer (2P50CA083639), a Program Project Development Grant from the Ovarian Cancer Research Fund, Inc., the Zarrow Foundation, the Marcus Foundation, and the Betty Ann Asche Murray Distinguished Professorship.

References

1. Elbashir, S. M., Lendeckel, W., Tuschl, T. (2001) RNA interference is mediated by 21- and 22-nucleotide RNAs. *Genes Dev.* **15**, 188–200.

2. Chendrimada, T. P., Gregory, R. I., Kumaraswamy, E., Norman, J., Cooch, N., Nishikura, K., et al. (2005) TRBP recruits the Dicer complex to Ago2 for microRNA processing and gene silencing. *Nature* **436**, 740–744.

3. Meister, G., Landthaler, M., Patkaniowska, A., Dorsett, Y., Teng, G., Tuschi, T. (2004) Human Argonaute2 mediates RNA cleavage targeted by miRNAs and siRNAs. *Mol. Cell.* **15**, 185–197.

4. Okamura, K., Ishizuka, A., Siomi, H., Siomi, M. C. (2004) Distinct roles for Argonaute proteins in small RNA-directed RNA cleavage pathways. *Genes Dev.* **18**, 1655–1666.

5. Wadhwa, R., Kaul, S. C., Miyagishi, M., Taira, K. (2004) Vectors for RNA interference. *Curr. Opin. Mol. Ther.* **6**, 367–372.

6. Ryther, R. C., Flynt, A. S., Phillips, J. A., Patton, J. G. (2005) SiRNA therapeutics: big potential from small RNAs. *Gene Ther.* **12**, 5–11.

7. Yu, D., Peng, P., Dharap, S. S., Wang, Y., Mehlig, M., Chandna, P., et al. (2005) Antitumor activity of poly(ethylene glycol)-camptothecin conjugate: the inhibition of tumor growth in vivo. *J. Control Rel.* **110**, 90–102.

8. Fang, J., Sawa, T., Maeda, H. (2003) Factors and mechanism of "EPR" effect and the enhanced antitumor effects of macromolecular drugs including SMANCS. *Adv. Exp. Med. Biol.* **519**, 29–49.

9. Greish, K., Fang, J., Inutsuka, T., Nagamitsu, A., Maeda, H. (2003) Macromolecular therapeutics: advantages and prospects with special emphasis on solid tumor targeting. *Clin. Pharmacokinet.* **42**, 1089–1105.

10. Maeda, H. (2001) The enhanced permeability and retention (EPR) effect in tumor vasculature: the key role of tumor-selective macromolecular drug targeting. *Adv. Enzyme Regul.* **41**, 189–207.

11. Matsumura, Y. and Maeda, H. (1986) A new concept for macromolecular therapeutics in cancer chemotherapy: mechanism of tumoritropic accumulation of proteins and the antitumor agent smancs. *Cancer Res.* **46**, 6387–6392.

12. Miller, C. R., Bondurant, B., McLean, S. D., McGovern, K. A., O'Brien, D. F. (1998) Liposome–cell interactions in vitro: effect of liposome surface charge on the binding and endocytosis of conventional and sterically stabilized liposomes. *Biochemistry* **37**, 12875–12883.

13. Thaker, P. H., Deavers, M., Celestino, J., Thornton, A., Fletcher, M. S., Landen, C. N., et al. (2004) EphA2 expression is associated with aggressive features in ovarian carcinoma. *Clin Cancer Res.* **10**, 5145–5150.

14. Landen, C. N., Kinch, M. S., Sood, A. K. (2005) EphA2 as a target for ovarian cancer therapy. *Expert Opin. Ther. Targets* **9**, 51179–1187.

15. Landen, C. N., Merritt, W. M., Mangala, L. S., Sanguino, A. M., Bucana, C., Lu, C., et al. (2006) Intraperitoneal delivery of liposomal siRNA for therapy of advanced ovarian cancer. *Cancer Biol. Ther.* **5**, 1708–1713.

16. Landen, C. N., Chavez-Reyes, A., Bucana, C., Schmandt, R., Deavers, M. T., Lopez-Berestein, et al. (2005) Therapeutic EphA2 gene targeting in vivo using neutral liposomal small interfering RNA delivery. *Cancer Res.* **65**, 6910–6918.

17. Killion, J. J., Radinsky, R., Fidler, I. J. (1998) Orthotopic models are necessary to predict therapy of transplantable tumors in mice. *Cancer Metastasis Rev.* **17**, 279–284.

18. Voskoglou-Nomikos, T., Pater, J. L., Seymour, L. (2003) Clinical predictive value of the in vitro cell line, human xenograft, and mouse allograft preclinical cancer models. *Clin. Cancer Res.* **9**, 4227–4239.

Chapter 4

Therapeutic Applications of RNAi for Silencing Virus Replication

Ralph A. Tripp and Stephen Mark Tompkins

Abstract

RNA interference (RNAi) is an evolutionarily conserved gene-silencing mechanism in which small 19–23-nucleotide double-stranded RNA molecules, or small interfering RNAs (siRNAs), target cognate RNA for destruction with exquisite potency and selectivity. The RNAi machinery is believed to be expressed in all eukaryotic cells and has been shown to regulate host gene expression. Given this ability, RNAi silencing strategies have been developed to inhibit viral genes and replication in host cells. One area of growing interest is the development of synthetic siRNA drugs to target acute viral infections in which long-term gene silencing is not required or desirable. To achieve synthetic siRNA drug efficacy, these anti-viral agents need to be delivered to the appropriate host cells, as they do not readily cross the cell membrane. Varied delivery and siRNA chemical stabilization strategies are being investigated for siRNA drug delivery; however, several studies have shown that naked, unmodified siRNA drugs can be effective in silencing replication of some viruses in animal models of infection. These findings suggest that RNAi-based drugs may offer breakthrough technology to protect and treat humans and animals from viral infection. However, there are four major considerations for evaluating successful RNAi efficacy: the siRNAs must have high efficiency, show low cytotoxicity, result in minimal off-target effects, and lead to results that are reproducible between experiments. The methods and caveats to achieve these goals are discussed.

Key words: RNA interference, RNAi, small interfering RNA, siRNA, transfection, drug delivery, virus, prophylactic, therapeutic.

1. Introduction

RNA interference (RNAi) has been shown to be an important mechanism to specifically silence or downregulate gene expression. Key to the RNAi activity are small interfering RNA

John F. Reidhaar-Olson and Cristina M. Rondinone (eds.), *Therapeutic Applications of RNAi: Methods and Protocols*, vol. 555
© Humana Press, a part of Springer Science+Business Media, LLC 2009
DOI 10.1007/978-1-60327-295-7_4 Springerprotocols.com

(siRNA) strands having complementary nucleotide sequences to cognate RNA targets. The highly selective and robust activity of siRNAs on gene expression has made RNAi a valuable tool for investigating gene function and regulating gene activity. Since synthetic siRNAs can be readily made and introduced into cells to mediate suppression of specific genes of interest, this technology is now routinely used to study gene function in cultured mammalian cells (1), and to identify and validate potential drug targets in the host. However, RNAi has not been fully exploited as an anti-viral method, particularly for in vivo applications.

Use of RNAi in vivo has shown promise, and two basic methods for mediating RNAi have been used: delivery of siRNA (2, 3), and delivery of plasmid and viral vectors that express short hairpin RNAs (shRNAs) that are processed in situ into active siRNAs (2, 4). The field has predominantly focused on delivering siRNA because efficacy can be achieved by delivery across the cell membrane, whereas delivery across the nuclear membrane is not required, as it is for shRNA efficacy. Also, construction of shRNA expression vectors can be cumbersome, and substantially more time is required to test the efficacy of shRNA sequences. However, the success of siRNA-mediated gene silencing in vivo depends on delivery and maintenance of intact siRNA in the cells. Advances in this technology will be required to allow for highly efficient and specific siRNA delivery. Improvements in delivery methods will no doubt enhance the utility of the technique as a platform drug for disease intervention. Some progress has been made in siRNA delivery strategies that center on three general strategies: first, the development of novel siRNA conjugations and/or formulations to specifically target siRNA drugs; second, the development of siRNA carrier and/or encapsulation materials to protect siRNA drugs; and finally, optimal drug delivery methods, such as intravenous, aerosol, and topical (2, 3). Given the drug-like properties of siRNA, and recent studies showing that naked or chemically unmodified siRNA duplexes may have robust virus gene-silencing activity in vitro and in vivo, there is good rationale and strong promise for therapeutic applications of RNAi for silencing virus replication.

2. Materials

2.1. Selection of siRNA Sequences for Targeting of mRNAs

1. The design of siRNA duplexes for targeting a specific gene requires knowledge of the target sequence. The decision to target specific viral genes may depend on factors such as

whether those genes code for receptors or affect the anti-viral immune response; however, a conservative approach would be to target transcripts from highly conserved genes critical in development of the virus nucleocapsid or involved in transcription of the virus genome, e.g., polymerase. For additional information refer to **Notes 1–13**.

2.2. Annealing of siRNAs to Produce Duplexes

1. Many commercial manufacturers, such as Dharmacon-ThermoFisher (Lafayette, CO), offer a range of pre-synthesized siRNA duplexes to meet the needs of the investigator; however, custom RNA synthesis may require an annealing step prior to testing.

2. To begin, prepare 2X annealing buffer (200 mM potassium acetate, 4 mM magnesium acetate, 60 mM HEPES–KOH, pH 7.4).

3. Prepare a 20 μM siRNA duplex working stock by combining 2X annealing buffer with sense oligonucleotide (20 μM final concentration) and antisense oligonucleotide (20 μM final concentration) in sterile water.

4. Incubate for 1 min at 90°C followed by 1 h at 37°C, and store the working stock solution at −20°C. The siRNA duplex solution can be freeze-thawed as required, but all RNA solutions should be kept on ice to reduce hydrolysis.

5. For quality control, the duplex solution should be run on a low-melt agarose gel in 0.5X TBE buffer at 80 V for 1 h, and the RNA bands subsequently visualized under UV light after ethidium bromide staining.

2.3. Selection of siRNA Transfection Tools

1. There are several approaches to RNAi but two are common for siRNA delivery: (a) lipid-mediated transfection of synthetic siRNA, and (b) viral-mediated transduction of siRNA. Determining which one of these approaches to use depends on the cell type studied and whether transient or stable knockdown is desired. For additional information refer to **Notes 14–25**.

2.4. Selection of Cell Lines for siRNA Screening

1. Various mammalian cell lines can be used for screening siRNA candidates, but Vero cells (ATCC No. CCL-81) or Vero E6 cells (ATCC No. CRL-1586) are preferred for reasons discussed in **Notes 26–29**.

2.5. SDS-PAGE Electrophoresis Reagents

1. 10X running buffer: 30 g of Tris base; 140 g of glycine; 10 g of SDS, all diluted in 1 l of distilled water.

2. Stacking gel buffer: 0.5 M Tris-HCL, pH 6.8; 0.4% SDS.

3. 2X solubilization buffer: 60 mM Tris-HCl, pH 6.8; 20% SDS; 0.0004% bromophenol blue; 10% mercaptoethanol; 20% glycerol.

4. Separation gel buffer: 1.5 M Tris-HCL, pH 8.8; 0.4% SDS.

5. 10X transfer buffer: 30.3 g Tris base; 144 g glycine, all diluted in 1 l distilled water; adjust pH with HCl to 8.3.

6. 10X TBS-T buffer: 24.2 g Tris base; 80 g of NaCl, pH 7.5; 10 ml of Tween-20, all diluted in 1 l of distilled water. Store at 4°C.

3. Methods

3.1. Evaluating siRNA Activity in Cells

1. Before siRNAs are applied, it may be important to establish if the cells to be studied are susceptible to RNAi, as it is possible that some cell lines have lost their ability to perform RNAi. This can be done using commercial reporter plasmids encoding firefly (*Photinus pyralis*) luciferase, GFP, or similar reporter molecules. Refer to **Notes 30–31** for additional cell type information.

2. To quantify siRNA concentrations that are unknown, the investigator should use Beer's Law: (Absorbance at 260 nm) = (epsilon)(concentration)(path length in cm), where epsilon is the molar extinction coefficient.

3. When solved for the unknown, the equation becomes: Concentration = (Absorbance at 260 nm)/[(epsilon)(path length in cm)]. When a standard 10 mm cuvette is used, the path length variable in this equation is 1 cm. If a different size of cuvette is used, e.g., a 2 mm microcuvette, then the path length variable is 0.2 cm.

4. Use the molecular weight of siRNA to convert between nmol and μg. If the specific molecular weight is not known, you may use the average molecular weight of siRNA, which is 13,300 g/mol. Therefore, 1 nmol siRNA = 13.3 μg, or 1 mg siRNA = 75 nmol.

5. In determining the quantity of siRNA that should be screened in an in vitro experiment, it is best to include a final concentration of range between 50–100 nM. For more information regarding preliminary in vitro screening refer to **Notes 32–37.**

6. Vero cells are ideal to begin in vitro studies. Vero cells should be maintained at the proper CO_2 in a humidified incubator at 37°C and grown in the appropriate media, e.g., Dulbecco's modified Eagle's medium with 10% fetal bovine serum (FBS), but without antibiotics.

7. Sub-culturing is performed by passage using 0.25% (w/v) Trypsin in 0.53 mM EDTA solution using a sub-cultivation ratio of 1:3 with medium renewal twice per week.

8. For propagation and maintenance, warm the DMEM media and Trypsin-EDTA in a 37°C water bath

(Trypsin-EDTA made by diluting the stock 1:10 by adding PBS only).

9. Split the cells at 80–90% confluency by first decanting the media and washing the cells with PBS.

10. Aspirate off the PBS, add Trypsin-EDTA, and incubate at 37°C for approximately 3 min – be careful not to over incubate with Trypsin-EDTA as it will damage the cells.

11. Remove cells from the incubator and loosen the cells by gentle aspiration with a sterile pipette.

12. Place the dissociated cells in a sterile plastic tube with media containing 1% FBS to aid inactivation of the trypsin.

13. Count the cells using a hemacytometer or related device.

14. Dilute the cells in DMEM containing 10% FBS to an appropriate cell number for the vessel that will propagate them, usually about 1:5.

15. Repeat this procedure every 3–4 days, maintaining them twice a week so that they are not diluted too much or overgrown.

16. Refer to **Notes 38** and **39** regarding Vero cell storage and recovery.

17. For siRNA studies, 24 h prior to siRNA transfection, the cells should be trypsinized, washed in fresh DMEM medium without antibiotics, and aliquots transferred into each well of a 24-well plate. The Vero cells should be plated in 500 μl of medium without antibiotics so that that they will be approximately 50–70% confluent at the time of transfection.

18. For each transfection, prepare optimal siRNA/lipid transfection reagent complex as per the manufacturer's instructions. A dose–response should be performed with siRNA candidates and siRNA controls at concentrations ranging from 1 to 50 nM. Refer to **Note 40** for additional information.

19. Add the complex to the cells and gently mix by rocking the plate.

20. Incubate the plate at 37°C in a CO_2 incubator for 12–24 h. Transfection times may vary depending on siRNA stability, the siRNA sequence chosen, and the growth characteristics of the cell line being transfected.

21. If cell viability loss occurs, the transfection reagent–siRNA complex medium may be changed after 6–8 h.

22. Following transfection, the activity should be examined between 24 and 72 h siRNA. This can be done using a reporter-based assay (refer to **Notes 18** and **29**) that may be suitable for quantification by fluorescence-activated cell sorting, fluorescence, or as recommended by the manufacturer. It is important to note that transfection reagents change the autofluorescence of

cells, which must be controlled for by mock transfection.

23. For evaluating endpoints of siRNA activity in cells, refer to **Notes 41–50**.

3.2. Testing Viral Endpoints Following siRNA Treatment

There are several assays to evaluate siRNA efficacy against viruses, which fall into two main strategies, prophylactic or therapeutic application. Refer to **Note 51** for general information.

3.2.1. Virus Plaque Assay for Prophylactic siRNA Treatment Against Respiratory Syncytial Virus (RSV)

1. 24 h prior to siRNA testing, 24-well plates of the cell line of interest (e.g., Vero E6 cells) must be seeded to obtain 80–90% confluency.
2. Transfect the cells using a commercial transfection reagent, such as Mirus TransIT-TKO (TKO). Follow the manufacturer's transfection guidelines.
3. For example, for TKO-mediated transfection, add 2 μl of TKO to 50 μl of DMEM cell media that does not contain FBS and incubate at room temperature for 10 min.
4. Add siRNA (final optimal concentration needed is often ~100 nM) to TKO/DMEM mixture and incubate at room temperature for 10 min.
5. Remove tissue culture media from 24-h cultures of Vero E6 cells.
6. Wash gently with PBS and decant.
7. Add the siRNA/TKO mixture to cells and gently rock for 3 min.
8. Be sure to include controls: TKO only and scrambled or mismatched siRNA/TKO complex.
9. Incubate for 12 h (overnight) at 37°C, 5% CO_2.
10. At between 1–2 h post-siRNA transfection, remove plate from incubator and remove siRNA mixture from cells.
11. Prepare RSV virus (100–200 PFU/100 μl) in DMEM without FBS.
12. Add 100 μl of diluted RSV to appropriate wells of 24-well plate (test all siRNA concentrations in triplicate) and incubate for 1 h at 37°C, 5% CO_2.
13. At 2 h post-infection, remove plate from incubator and add 1 ml of DMEM/5% FBS containing 2% methylcellulose media to each well and incubate plate at 37°C, 5% CO_2 for 4–5 days.
14. At day 4 or 5 post-infection, remove overlay and visualize plaques by immunostaining with antibodies reactive to RSV F and/or G proteins, or counterstain, e.g., neutral red dye. Refer to **Note 52** for a word of caution about the assay.
15. To evaluate results, determine the percent plaque reduction from the controls, i.e., mismatched siRNA control and TKO control.

3.2.2. Reverse Transcriptase-Polymerase Chain Reaction (RT-PCR) and Quantitative Real-Time PCR (Q-PCR)

When using these methods to evaluate siRNA silencing of a gene, it is important to consider the interdependence between the targeting siRNA sequence and the PCR amplification region for assessing the efficacy of the siRNA target gene knockdown. Refer to **Note 53** for general information.

A generalized method for RT-PCR and Q-PCR is given below:

1. Total RNA is extracted from the transfected cells with RNeasy Mini Kit (QIAGEN).
2. Total RNA (2 μg) is reverse-transcribed to single-stranded cDNA using an oligo(dT) primer and ThermoScript RNase H$^-$ reverse transcriptase (Invitrogen) according to the manufacturer's instructions.
3. Specific primers for the gene of interest must be prepared.
4. The RT-PCR reaction is carried out using the following mixture: 0.4 μl of cDNA template, 0.2 μl each of 1 mM gene-specific primers, 1.6 μl of 10 mM dNTPs, 2 U (0.4 μl) of Ex *Taq* (Takara Bio Inc., Siga, Japan), and 2 μl of 10 × PCR buffer.
5. Generally, 25 cycles of amplification are performed under the following conditions: 96°C for 30 s (denaturation), 30 s at 55°C (annealing), and 30 s at 72°C (extension). The PCR products are separated by electrophoresis in a 1.0% agarose gel containing 0.5 μg/ml ethidium bromide.
6. Q-PCR is carried out on a Real-Time PCR Detection System often using a QuantiTect SYBR Green PCR kit (QIAGEN). The Q-PCR reaction is carried out using the following mixture: 0.1 μl of cDNA template, 2 μl each of 1 mM specific primers, 25 μl of QuantiTect SYBR Green PCR Master Mix.
7. The Q-PCR amplification is performed under the following conditions: denaturation at 95°C for 15 min followed by 40 cycles of denaturation at 94°C for 30 s, annealing at 57°C for 30 s, and extension at 72°C for 30 s.
8. After Q-PCR, a melting curve is constructed by increasing the temperature from 55°C to 95°C.
9. Each sample for three independent experiments is run in quadruplicate. The mean cycle threshold (C_t) values for the gene are calculated and normalized to those for beta-actin. The mean value derived from each three experiments is calculated, and the results graphed with the corresponding standard deviation.

3.2.3. Western Blot Analysis

This is a method to detect a specific protein in the cell lysate using gel electrophoresis to separate native or denatured proteins by the length of the polypeptide (denaturing conditions) or by the 3-D structure of the protein (native/non-denaturing conditions). The proteins are then transferred to nitrocellulose or a

PVDF membrane where the protein can be detected, generally using antibodies. A general method for Western blotting is given below:

1. SDS-PAGE electrophoresis reagents are prepared as indicated.
2. To prepare 10% separation gel, mix 2.5 ml of 40% acrylamide/bis (9:1) solution, 2.5 ml of separation gel buffer, 5 ml of distilled water, 50 μl of ammonium persulfate, and 5 μl of TEMED. Cover the polymerizing gel by the layer of distilled water as it solidifies in the rig.
3. To prepare stacking gel, mix 1 ml of 40% acrylamide/bis (9:1) solution, 2.5 ml of stacking gel buffer, 6.5 ml of distilled water, 50 μl of ammonium persulfate, and 10 μl of TEMED. Dump out the layer of distilled water and layer the stacking gel on top of the separating gel.
4. To prepare sample, homogenize the particular material in solubilization buffer and incubate 30 min to 1 h at 40°C. Centrifuge sample for 2 min at 12,000 g.
5. Start electrophoresis in the pre-cooled running buffer at 70 V until the sample enters the separating gel, and then increase voltage up to 100 V.
6. To transfer the separated polypeptides onto the blotting paper, soak the blotting paper, e.g., Immobilon P cut to the size of the separating gel, in methanol for 2 min, and then accurately put the paper on the surface of water for 10 min. During this procedure the water replaces methanol from the paper pores.
7. Mix 100 ml of 10X transfer buffer with 850 ml of distilled water, cool the solution in the cold room, and add 50 ml of methanol.
8. Cut off the stacking gel and soak the separating gel in the prepared solution for 10 min at 4°C.
9. Soak two pieces of the Whatman 3MM paper with a size of separation gel in the prepared solution for 2 min.
10. Assemble sandwich, remove bubbles between gel and Immobilon P, and perform electrotransfer in the prepared solution for 1 h at 100 V.
11. Prepare 100 ml of TBS-T buffer. Add 5 g of dry nonfat milk. Shake vigorously.
12. Incubate the blotting paper in 15 ml of the prepared milk solution for 1 h at room temperature.
13. Wash the blotting paper two times with 15 ml of TBS-T buffer for 5 min.
14. Dissolve the primary antibodies in 15 ml of milk TBS-T buffer and add to the blotting paper. Incubate 1 h at room temperature.
15. Wash the paper with 15 ml of TBS-T buffer four times for 5 min.

16. Dissolve the secondary antibodies, e.g., conjugated to horseradish peroxidase, in 15 ml of milk TBS-T buffer and add to the blotting paper.

17. Prepare 6 ml of the substrate solution for horseradish peroxidase in accordance with manufacturer's protocol.

18. Incubate the blotting paper with this solution for 1 min. Wrap the wet paper with the Cling Wrap, put the surface of the covered blotting paper on film or imager, and detect.

3.2.4. Cell Cytotoxicity

Transfection-mediated cell cytotoxicity may be observed particularly during long-term exposure to transfection reagents. For general information refer to **Note 54**. The general protocol for performing cell cytotoxicity assays using the WST-1 method *(5)*, a highly sensitive and easy method, is given below:

1. WST-1 reagent is added to supernatants collected from transfected and control untransfected cell lines, i.e., 10 μl/100 μl of cell supernatant. This should be done in a 96-well plate format suitable for an ELISA plate reader.

2. The mixture is incubated for 0.5 h, and then the absorbance measured at wavelength 420–480 nm. Analyze the absorbance values to determine cell cytotoxicity compared to control wells.

3. Refer to **Note 55** for information regarding other commercial cytotoxicity kits.

3.3. Potential Pitfalls

There are important pitfalls to consider that may be encountered in prophylactic or therapeutic applications of RNAi. These include situations encountered in designing siRNA candidates, the host cell response to siRNAs, and evaluating efficacy. Refer to **Notes 56–64** for more information.

4. Notes

1. In considering target sites, intronic sequences should be avoided. Sequence information about mature mRNAs can be located from EST databases, e.g., www.ncbi.nlm.nih.gov.

2. It is important to compare the potential target sites to the appropriate genome database (examples given herein) to eliminate siRNA candidates with target sequences having more than 16–17 contiguous base pairs of homology to other coding sequences. Upon deciding on the target, the AUG start codon of the transcript should be scanned for AA dinucleotide sequences and the 3′ adjacent 19 nucleotides as potential siRNA target sites.

3. Since regions of mRNA may be highly structured or capped with regulatory proteins, several different siRNA target sites should be selected at different positions along the length of the gene sequence to ensure efficacy. It is important to note that the siRNA specificity for the target region is critical as a single point mutation is sufficient to abolish target mRNA degradation (6).

4. Target sequences can be predicted from genomic sequences using prediction programs, e.g., exon.gatech.edu/GeneMark, genes.mit.edu/GENSCAN.html, or others.

5. In thinking about siRNA design, it is important to note that the most predominantly naturally processed and effective siRNAs have been shown to be duplexes of 21- and 22-nt with symmetric 2-nt 3′ overhangs (6, 7).

6. Standard selection rules for siRNA design include the following: (**a**) the targeted regions on the mRNA sequence should be located 50–100 nt downstream of the start codon (ATG); (**b**) one should search for the sequence motif $AA(N_{19})TT$ or $NA(N_{21})$, or $NAR(N_{17})YNN$, where N is any nucleotide, R is purine (A, G), and Y is pyrimidine (C, U); (**c**) avoid sequences with >50% G+C content, and avoid stretches of four or more nucleotide repeats; (**d**) generally avoid targeting 5′ UTR and 3′ UTR regions; and (**e**) avoid sequences that share a certain degree of homology with other related or unrelated genes (7).

7. Many commercial sites exist where one can send in the target gene information, and, using proprietary algorithms, the companies will develop synthetic siRNAs or siRNA pools for testing.

8. The siRNAs chosen for commercial design should be 19–23 nt in length and chemically synthesized using ribonucleoside phosphoramidites.

9. A typical 0.2 μmol-scale RNA synthesis generally provides about 1 mg, sufficient for approximately 1,000 in vitro transfection experiments using a 24-well tissue culture plate format.

10. There are a number of siRNA synthesis options including water-soluble, 2′-O-protected RNA which can be deprotected in aqueous buffers prior to use; fully deprotected RNA; and purified siRNAs duplexes. For most applications, the third option is preferred.

11. It is critical that the siRNA-targeted region on the mRNA sequence does not share significant, if any, homology with other genes or sequences in the genome, in order to minimize off-target effects.

12. It is preferable to select target regions such that siRNA sequences may contain uridine residues in the 2-nt

overhangs that can be replaced by 2′-deoxythymidine without loss of activity. This feature reduces costs of RNA synthesis and appears to enhance nuclease resistance.

13. It is advisable to synthesize siRNA specificity controls by scrambling the sequence of the effective siRNA duplex.

14. The most popular application for transient transfection of siRNA is the use of cationic lipid-based reagents because they are generally effective for delivering siRNA in many commonly used cell lines. There are many commercial suppliers of lipid-based transfection reagents that include Lipofectamine 2000™ (Invitrogen Corporation Carlsbad, CA) and Transit-TKO™ (Mirus Bio Corporation, Madison, WI).

15. Although lipid-based transfection is one of the more commonly used methods for adherent cells, suspension cells are often more difficult to transfect and generally have higher rates of delivery with electroporation techniques.

16. Although siRNA transfection using lipid-based reagents is a transient event, maximal efficacy is typically achieved 24–48 h after transfection, and may last for hours to days depending on the efficiency of transfection and the cell type being treated. A typical transfection protocol includes plating the cells, creating transfection complexes by mixing transfection agent with siRNA, and adding the transfection mixture to the growth medium with the cells.

17. shRNA can be expressed from lentivirus vectors allowing for high-efficiency siRNA transfection of a variety of cell types, including non-dividing cells and differentiated neurons of the brain (8, 9). Approaches for lentiviral hairpin-mediated RNAi are numerous, as are methodologies for deriving RNAi-transgenic cells.

18. There are a multitude of commercially available lentivirus RNAi vectors, available at various websites, e.g., www.systembio.com, www. dharmacon. com, www.genscript.com, www. invitrogen.com, and others. Many of these vectors encode green fluorescent protein (GFP) or other markers for assessing infectivity, while other vectors are available with antibiotic selection markers such as puromycin. In addition, there are conditional lentiviral RNAi vectors available, with which the user can turn silencing on or off as required during the course of experimentation.

19. If a lentivirus RNAi vector must be constructed, an important first step is to design an effective RNA hairpin construct. Rules have been developed to enrich for successful hairpin constructs based on thermodynamic

analysis of siRNA efficacy. There are several prediction algorithms available for designing hairpin RNA, e.g., www-lbit.iro.umontreal.ca/mcfold/index.html and others, and several institutions have created lentiviral RNAi libraries that are available, e.g., mcmanuslab.ucsf.edu/.

20. Creating hairpin RNAi vectors can be accomplished by different methods. One approach involves annealing two long complementary DNA oligomers that are directionally cloned into the lentivirus expression vector. In a more involved approach, the entire hairpin sequence is included as part of one of the oligomers where the 300-nt U6 promoter is to be amplified by PCR and the resulting product contains the hairpin entity. These two approaches require long oligomers, thus often the final constructs are subject to sequence errors. To overcome this, other methods have been developed such as the synthesis of four different DNA oligomers whereby the oligomers are annealed and directly cloned into the lentivirus vector.

21. Since off-targeting of siRNAs generated by the lentivirus vectors may occur, it is advised that at least two different hairpin vectors should be made for each targeted gene. Thus, the vectors could be tested to determine if they have the same phenotypes.

22. To prepare a competent lentivirus from the DNA vectors, the DNA vectors are transiently transfected into a packaging cell line such as human 293 cells, and after 2–3 days the supernatant will contain the virus. These production systems generally incorporate a spilt-system, where the lentiviral genome has been split into individual helper plasmid constructs which diminishes the risk of creating a replication-capable virus by adventitious recombination of the lentiviral genome.

23. It is important to consider if the lentiviral vectors being prepared should have a restricted or broad host range. This feature is generally determined by the pseudotype or virus coat used – a feature that can be easily regulated as the lentiviral production systems are split. To generate a broadly tropic vector, one may consider using the vesicular stomatitis virus (VSV) glycoprotein, as opposed to another binding surface protein which would display limited host cell specificity.

24. After transfection of the 293 cells or other cell types has been achieved, the virus in the supernatant may be separated from the cells by filtration of the supernatant through a 0.45-μm syringe-filter. It is important to determine the virus titer in the supernatant so that experiments can be reproduced over the course of multiple experiments.

25. Different cell types differ in their ability to be infected. Polybrene (hexadimethrine bromide), a cationic polymer, may be used to increase the efficiency of infection of certain cells, as it acts by neutralizing the charge repulsion between the virions and cell surface.

26. As some RNA duplexes may stimulate type I IFN expression upon transfection into certain cell types, it is important to consider if a type I IFN-defective cell line can be used in the screening process, such as Vero or Vero E6 cells, a continuous line of African green monkey kidney cells defective in the production of interferon. This feature is particularly important for screening siRNA candidates that target viruses, as induction of type I IFNs can inhibit virus replication *(10–12)*.

27. It is important to maintain healthy cells. Some cell lines are more sensitive to transfection agents than others, and this may be dependent upon passage number. It is advisable to use cells subjected to a similar number of passages to ensure reproducibility in transfection results between experiments.

28. It is recommended that control transfections be performed by varying cell confluency (40–90%), as low cell density or too much transfection reagent increases cell toxicity.

29. To estimate the transfection efficiency in a particular cell line, it is recommended that fluorophore-conjugated siRNA be tested prior to experimentation, e.g., Cy3-labeled siRNA or fluorescein-siRNA transfection controls.

30. For studies involving siRNAs targeting viral genes it is often best to screen in vitro using flat-bottom 24-well plates as this allows for enumeration of virus titers as an endpoint.

31. For these type of studies, Vero or Vero E6 cells are preferred, but other cell lines have been used, including HeLa, COS, NIH/3T3, HEK 293, CHO, A549, and Drosophila cells *(13–15)*.

32. In preliminary in vitro siRNA screenings, the cells should be transfected as indicated by the manufacturer using a reliable transfection agent, e.g., Lipofectamine 2000 (Invitrogen).

33. The optimal ratio of siRNA-to-lipid transfection reagent and the total amount of this complex should be determined using the same number of Vero cells to evaluate the highest transfection efficiency.

34. Optimal ratios should be approached by starting with a constant amount of siRNA and varying the amount of transfection reagent. The best place to start is following the manufacturer's recommendations as each transfection reagent is different.

35. Reduced-serum medium is recommended for transfection; however, no antibiotics should be added to the media during transfection.

36. It is important to maintain the same cell numbers and conditions across experiments.

37. The serum-free media to be used in the assay should be tested for compatibility with the transfection reagent. Opti-MEM® is generally a good media to start trials with.

38. It is best not to exceed a passage number of 30 after thawing the stock culture, so make a master freezing of low passaged cells as the number of passages may affect siRNA transfection efficiencies.

39. Aliquots of cells of low passage number cells may be stored frozen by centrifuging the cells in a sterile tube and resuspending in a 9:1 ratio of FBS:DMSO. The cells should be aliquoted into appropriate freezer vials, put into Mr. Frosty containers first for freezing at $-70°C$ and then moved into liquid nitrogen.

40. To transfect the cells with the siRNA candidates, it is best to use the conditions recommended by the manufacturer for the transfection reagents being used as they operate differently. For example, INTERFERin™, DharmaFECT™, HiPerFect™, Transit-TKO™, Lipofectamine 2000™, and other transfection reagents are available for siRNA transfection of eukaryotic cells, especially at low siRNA concentrations, e.g., 1–10 nM siRNA concentrations.

41. Optimally, siRNA should be delivered to the cells 12–24 h prior to virus infection for siRNA drugs targeting viruses. Endpoints such as virus titers are generally assessed 48 h post-transfection. It is important not to overwhelm the cells during the infection phase; multiplicities of infection (MOIs) between 0.1 and 1.0 are generally used.

42. siRNAs targeting different positions on the same gene or different genes may induce different levels of gene silencing. Therefore, the selection of the mRNA target sequence is critical to siRNA efficacy. Equally important are the endpoints for analysis of siRNA efficacy. For virus studies these endpoints may include examining virus titer, qPCR of virus gene expression, plaque morphology, viral antigen levels, or loss of reporter signal such as that associated with a GFP-virus.

43. Effective siRNA knockdown of virus replication can be deduced by determining the reduction of virus titer, or at least demonstrating a reduction of the targeted viral mRNA. An immunostaining virus plaque assay using monoclonal antibodies specific to a viral protein will allow for quantitation of virus replication.

44. Standard plaque assays using counterstaining may also be effective, but for this and immunostaining plaque assays, it is recommended that a commercial plaque counting device be used to minimize investigator error. ELISPOT plate readers such as the AID EliSpot Reader has a software program that counts spots according to the user's settings (www.elispot. com/index.html?elispot_reader/software.htm).

45. It may be useful to examine expression of the targeted viral protein by Western blotting of virally infected cell lysates.

46. If no siRNA knockdown of protein is observed by Western blot, it may be desirable to analyze whether the target mRNA was effectively destroyed by the transfected siRNA. This can be tested by preparing total RNA from the transfected and mock-treated controls between 24 and 72 h post-transfection.

47. The total RNA can be reverse-transcribed using a target-specific primer and PCR-amplified with a primer pair covering at least one exon–exon junction in order to control for amplification of pre-mRNAs.

48. RT-PCR of a non-targeted mRNA is also needed as control. Effective depletion of the mRNA with undetectable reduction of target protein may indicate that a large reservoir of stable protein exists in the cell. Multiple transfections in sufficiently long intervals may be necessary until the target protein is finally depleted to a point where a phenotype may become apparent.

49. If multiple transfection steps are required, it is recommended that the cells be split every 2–3 days after transfection. The cells may be transfected immediately after splitting. Note that cells diluted to a confluency of less than 60% are generally less effectively transfected.

50. If detection of siRNA-mediated knockdown cannot be observed, test a different siRNA duplex targeting a different gene region or gene, test for the possibility of sequencing errors in deposited sequence files, or for polymorphisms that may be a problem for a given sequence. It is recommended that that the cell line used be investigated, as ~36% of cell lines are of different origin or species to that claimed *(16)*.

51. Prophylactic treatment is the most common method used to evaluate efficacy, where the siRNAs are delivered to the cells of choice prior to virus infection. Therapeutic treatment is done after virus infection of the cell line. In both cases, virus plaque assays are commonly used to evaluate efficacy. A plaque assay measures the number of infectious virus particles. The methods for these assays vary and are dependent upon the virus being studied, as viruses

exhibit cell tropism, cell receptor requirements, cell enzyme requirements, etc. It is important to be familiarized with the current literature for the appropriate methods to use for the virus to be studied.

52. A word of caution – HEp-2 cells (ATCC # CCL-23) are also commonly used to propagate and titer RSV. It is important to note that this line was originally thought to be derived from an epidermoid carcinoma of the human larynx, but was subsequently found based on isoenzyme analysis, HeLa marker chromosomes, and DNA finger-printing, to have been established via HeLa cell contamination. Thus, these cells are not recommended for use.

53. Accurate measurement of siRNA efficacy by PCR requires careful attention to the primer location used to amplify the target mRNA. It is known that initial cleavage of the target gene mRNA by the RNA-induced silencing complex (RISC) endonuclease occurs near the center of the siRNA targeting sequence (6), an effect mediated by Ago-2 activity (17); however, complete degradation of the entire mRNA is not always observed (18, 19).

54. Some siRNAs have been reported to induce off-target effects linked to cell cytotoxicity, an effect that may obscure evaluation of siRNA efficacy. There are a number of commercially available kits for evaluating cell cytotoxicity. The WST-1 method (5) is highly sensitive and easy to perform. WST-1 is a tetrazolium salt that when in contact with metabolically active cells gets cleaved to formazan by mitochondrial dehydrogenases. The formazan dye is then measured using a scanning spectrophotometer at wavelengths 420–480 nm. The darker the formazan dye, the greater the number of metabolically active cells in that well. The WST-1 reagent has several advantages compared to other cell cytotoxicity reagents in the market. The WST-1 reagent is water-soluble after cleavage so there is no need to perform a solubilization step, and WST-1 is very stable so it can be stored in a ready-to-use solution.

55. There are also commercial kits that provide a direct measurement of cytotoxicity rather than using an indirect indicator such as the release of cell enzymes. An example of a kit is given by the LIVE/DEAD® Viability/Cytotoxicity Assay Kit from Molecular Probes. This kit provides a two-color fluorescence cell viability assay that is based on the simultaneous determination of live and dead cells with two probes that measure two recognized parameters of cell viability: intracellular esterase activity and plasma membrane integrity. The kit can be used with fluorescence microscopes, fluorescent plate scanners, or flow cytometers. The assay is applicable to most eukaryotic cell types, including

adherent cells. It is recommended that the user follow the manufacturer's protocol as it varies depending upon application.

56. When designing siRNA candidates it is important to note that if the number of allowed mismatches between siRNA and homologous gene sequences are set too stringently, some functional siRNAs will be rejected due to sequencing errors that are common to EST databases.

57. Several studies have shown that the host cell interferon-response effect is enhanced by increasing the concentration of siRNAs *(11, 12)*.

58. siRNA duplexes >23 bp can influence cell viability and induce a host cell IFN response that appears to be mediated through upregulation of the dsRNA receptor, Toll-like receptor 3 *(20)*. These effects may be cell type specific.

59. Several studies have shown that off-target effects may be caused by perfect matches of the seed region of the siRNA antisense strand with the 3′ untranslated region of unintended mRNA targets, combined with limited homology elsewhere in the 21-base target sequence *(21, 22)*.

60. Due to the risk of off-target effects, it is critical to design siRNA experiments with redundancy so that spurious results can be detected. Redundancy experiments should use multiple different silencing reagents, e.g., several siRNAs targeting different areas of the same mRNA. As the probability that several siRNAs that target different sequences of the same gene would cause the same phenotype through off-target interaction is very low, this approach can be used to show siRNA specificity and the validity of the relationship between target gene knockdown and resulting phenotype.

61. In designing short siRNAs, it is important to consider that they may target a part of the mRNA that is masked by secondary structure or bound proteins, thus reducing efficacy through inaccessibility of the target site.

62. A negative siRNA screening result may be due to inefficacy that could result from an inability to reduce translation to the point where a phenotype is apparent.

63. With regard to the therapeutic window of siRNA efficacy, it is important to consider the level of siRNA uptake and intracellular processing by the cell type being investigated.

64. If the goal is siRNA therapeutic approaches, it is important to consider the breadth of applicability to the clinically relevant target, the level of cell cytotoxicity associated with delivery of the siRNA, and other regulatory perspectives.

   Another limitation to consider is that clinical use of siRNAs would typically be transient in nature with intracellular siRNA concentrations and efficacy usually declining within a 1–2 week period. Thus, treatment of chronic pathologies would either require repetitive siRNA treatment, or long-term RNAi linked to stable transfection, e.g., lentivirus vectoring.

References

1. Elbashir, S. M., Harborth, J., Lendeckel, W., Yalcin, A., Weber, K., and Tuschl, T. (2001) Duplexes of 21-nucleotide RNAs mediate RNA interference in cultured mammalian cells *Nature* **411**, 494.

2. Amarzguioui, M., Rossi, J. J., and Kim, D. (2005) Approaches for chemically synthesized siRNA and vector-mediated RNAi *FEBS Lett* **579**, 5974–5981.

3. de Fougerolles, A., Vornlocher, H.-P., Maraganore, J., and Lieberman, J. (2007) Interfering with disease: a progress report on siRNA-based therapeutics *Nat Rev Drug Discov* **6**, 443–453.

4. Fewell, G. D., and Schmitt, K. (2006) Vector-based RNAi approaches for stable, inducible and genome-wide screens *Drug Discov Today* **11**, 975–982.

5. Hamasaki, K., Kogure, K., and Ohwada, K. (1996) A biological method for the quantitative measurement of tetrodotoxin (TTX): tissue culture bioassay in combination with a water-soluble tetrazolium salt. *Toxicon* **34**, 490–495.

6. Elbashir, S. M., Lendeckel, W., and Tuschl, T. (2001) RNA interference is mediated by 21- and 22-nucleotide RNAs *Genes Dev.* **15**, 188–200.

7. Birmingham, A., Anderson, E., Sullivan, K., Reynolds, A., Boese, Q., Leake, D., Karpilow, J., and Khvorova, A. (2007) A protocol for designing siRNAs with high functionality and specificity *Nat Protocols* **2**, 2068–2078.

8. Gonzalez-Alegre, P., Bode, N., Davidson, B. L., and Paulson, H. L. (2005) Silencing primary dystonia: lentiviral-mediated RNA interference therapy for DYT1 dystonia. *J Neurosci* **25**, 10502–10509.

9. Dann, C. (2007) New technology for an old favorite: lentiviral transgenesis and RNAi in rats. *Transgenic Res* **16**, 571–580.

10. Ge, Q., McManus, M. T., Nguyen, T., Shen, C. H., Sharp, P. A., Eisen, H. N., and Chen, J. (2003) RNA interference of influenza virus production by directly targeting mRNA for degradation and indirectly inhibiting all viral RNA transcription *Proc Natl Acad Sci USA* **100**, 2718–2723.

11. Sledz, C. A., Holko, M., de Veer, M. J., Silverman, R. H., and Williams, B. R. (2003) Activation of the interferon system by short-interfering RNAs. *Nat Cell Biol* **5**, 834–839.

12. Sledz, C. A., and Williams, B. R. G. (2004) RNA interference and double-stranded-RNA-activated pathways *Biochem Soc Trans* **32**, 952–956.

13. Perrimon, N., Friedman, A., Mathey-Prevot, B., and Eggert, U. S. (2007) Drug–target identification in Drosophila cells: combining high-throughout RNAi and small-molecule screens. *Drug Discov Today* **12**, 28–33.

14. Birmingham, A., Anderson, E. M., Reynolds, A., Ilsley-Tyree, D., Leake, D., Fedorov, Y., Baskerville, S., Maksimova, E., Robinson, K., Karpilow, J., Marshall, W. S., and Khvorova, A. (2006) 3′ UTR seed matches, but not overall identity, are associated with RNAi off-targets. *Nat Meth* **3**, 199–204.

15. Cullen, L. M., and Arndt, G. M. (2005) Genome-wide screening for gene function using RNAi in mammalian cells. *Immunol Cell Biol* **83**, 217–223.

16. Masters, J. R., Thomson, J. A., Daly-Burns, B., Reid, Y. A., Dirks, W. G., Packer, P., Toji, L. H., Ohno, T., Tanabe, H., Arlett, C. F., Kelland, L. R., Harrison, M., Virmani, A., Ward, T. H., Ayres, K. L., and Debenham, P. G. (2001) Short tandem repeat profiling provides an international reference standard for human cell lines *Proc Natl Acad Sci USA* **98**, 8012–8017.

17. Lingel, A., and Izaurralde, E. (2004) RNAi: finding the elusive endonuclease. *RNA* **10**, 1675–1679.

18. Koller, E., Propp, S., Murray, H., Lima, W., Bhat, B., Prakash, T. P., Allerson, C. R., Swayze, E. E., Marcusson, E. G., and Dean, N. M. (2006) Competition for RISC binding predicts in vitro potency of siRNA. *Nucl Acids Res* **34**, 4467–4476.

19. Vickers, T. A., Lima, W. F., Nichols, J. G., and Crooke, S. T. (2007) Reduced levels of Ago2 expression result in increased siRNA competition in mammalian cells *Nucl Acids Res.* **35**, 6598–6610.

20. Reynolds, A., Anderson, E. M., Vermeulen, A., Fedorov, Y., Robinson, K., Leake, D., Karpilow, J., Marshall, W. S., and Khvorova, A. (2006) Induction of the interferon response by siRNA is cell type- and duplex length-dependent. *RNA* **12**, 988–993.

21. Lin, X., Ruan, X., Anderson, M. G., McDowell, J. A., Kroeger, P. E., Fesik, S. W., and Shen, Y. (2005) siRNA-mediated off-target gene silencing triggered by a 7 nt complementation. *Nucl Acids Res* **33**, 4527–4535.

22. Jackson, A. L., Burchard, J., Schelter, J., Chau, B. N., Cleary, M., Lim, L., and Linsley, P. S. (2006) Widespread siRNA "off-target" transcript silencing mediated by seed region sequence complementarity. *RNA* **12**, 1179–1187.

Chapter 5

RNAi in the Malaria Vector, *Anopheles gambiae*

Flaminia Catteruccia and Elena A. Levashina

Abstract

Malaria is a disease that kills more than a million people each year in tropical and subtropical countries. The disease is caused by *Plasmodium* parasites and is transmitted to humans exclusively by mosquitoes of the genus *Anopheles*. The lack of functional approaches has hampered study of the biological networks that determine parasite transmission by the insect vector. The recent discovery of RNA interference and its adaptation to mosquitoes is now providing crucial tools for the dissection of vector–parasite interactions and for the analysis of aspects of mosquito biology influencing the vectorial capacity. Two RNAi approaches have been established in mosquitoes: transient gene silencing by direct injection of double-stranded RNA, and stable expression of hairpin RNAs from transgenes integrated in the genome. Here we describe these methods in detail, providing information about their use and limitations.

Key words: Malaria, mosquito, RNAi, transgenesis, immunoblotting, q-PCR.

1. Introduction

Human malaria persists today as one of the most widespread and devastating infectious diseases in the world. *Plasmodium* parasites are transmitted to humans when an infected *Anopheles* mosquito takes a blood meal. The relationship between the mosquito vector and the malaria parasite is shaped by a complex network of biological interactions determining whether a given mosquito species will be capable of sustaining parasite development. Moreover, the vector competence to transmit disease is also governed by a series of factors intrinsic to mosquito biology, such as host preferences, longevity, immunity, and reproductive rates. However, while efficient genetic methods to study gene function in *Plasmodium* have long been established *(1)*, functional studies in the

John F. Reidhaar-Olson and Cristina M. Rondinone (eds.), *Therapeutic Applications of RNAi: Methods and Protocols, vol. 555*
© Humana Press, a part of Springer Science+Business Media, LLC 2009
DOI 10.1007/978-1-60327-295-7_5 Springerprotocols.com

mosquito vector are lagging behind, and until recently little was known about the factors influencing *Plasmodium* development in the *Anopheles* vector.

Since the recent discovery of RNA interference, the "RNAi revolution" has reshaped the field of functional genomics, allowing characterization of genomes previously recalcitrant to targeted gene manipulation. This revolution has also hit mosquito research. Combined with the availability of the genome sequence of the major malaria vector, *Anopheles gambiae*, RNAi provides a new tool for malaria research, permitting study of molecules and mechanisms of mosquito biology that are relevant to disease transmission. Here, we describe two approaches for RNAi-based silencing in *Anopheles* for performing functional analysis in the mosquito vector. Further development of these methods might ultimately lead to new ways of controlling and perhaps even eradicating this devastating disease.

2. Materials

2.1. dsRNA Synthesis

1. pLL10 (**Fig. 5.1A**): A pBluescript-based plasmid with two T7 promoter sequences flanking the polylinker region in opposite directions *(2)*.
2. Proteinase K stock solution: 20 mg/mL in sterile 20 mM Tris (pH 8); 1.5 mM CaCl$_2$, 50% glycerol. Aliquots can be stored at −20°C. Proteinase K buffer: 10 mM Tris-HCl (pH 8); 10 mM EDTA (pH 8); 5 mM NaCl; 2 mM CaCl$_2$. Proteinase K final solution: add 1 μL of proteinase K (20 mg/mL) to 150 μL of buffer. Store in 50 μl aliquots (one tube for two reactions) at −20°C.
3. Linearized plasmid and RNA purification: RNase-free (DEPC-treated) water; phenol/chloroform/isoamyl alcohol (25:24:1); chloroform; isopropanol; 70% ethanol. All reagents should be RNase-free.
4. Synthesis and purification of single-stranded RNAs (ssRNAs): T7 MEGAscript kit (Ambion, Applied Biosystems, Foster City, CA, USA).

2.2. Plasmids for Stable RNAi and Injection Mixture

1. Transformation vector pBac[3xP3-EGFPafm] or similar, which contains a fluorescent protein selectable marker under control of the artificial *3xP3* promoter cloned within the inverted repeats of the *piggyBac* transposable element (**Fig. 5.1B**).
2. Helper plasmid phsp-pBac, which contains the piggyBac transposase gene driven by the *hsp70* promoter from *Drosophila melanogaster (3)* (**Fig. 5.1B**).

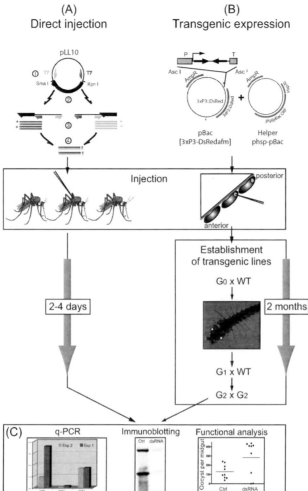

Fig. 5.1. **Methods for RNAi gene silencing in the mosquitoes**. (**A**) Direct injection of dsRNA. Map of pLL10 and position of the T7 promoters. The main steps in dsRNA synthesis are described: (1) cloning of the gene of interest in pLL10 between the two T7 promoters, using polylinker enzymes; (2) linearization of the plasmid on both sides of the insert; (3) synthesis of sense (+) and antisense (−) ssRNAs independently using the MEGAscript kit; (4) annealing of ssRNAs to form dsRNA (±). In vitro synthesized dsRNA is injected into the mosquito thorax (here shown for adults). Efficiency of dsRNA silencing can be estimated 12–24 h later by q-PCR or 2–4 days later by immunoblotting. The persistence of silencing often depends on the targeted gene, but in general it lasts for at least 5 days. (**B**) Stable RNAi gene silencing in transgenic mosquitoes. Cloning of the inverted fragments of the target gene into a *piggyBac* transformation vector. Injection of the transformation mixture, comprising the transformation vector and a transposase-expressing "helper" plasmid, into mosquito embryos. Expression of the selectable marker (in this case, expression of a red fluorescent protein is driven by the *Pax3* gene promoter in the eyes and nervous system (white)). Fluorescent G_1 individuals are outcrossed with wild-type mosquitoes (WT), and G_2 fluorescent progeny is intercrossed to amplify numbers of transgenic mosquitoes. Homozygous lines are then established by selection and intercross of homozygous individuals. (**C**) The efficiency of RNAi silencing in the injected and transgenic mosquitoes can be analyzed by q-PCR (transcriptional level) and immunoblotting (translational level). To date, a number of functional tests have been established to reveal the involvement of the targeted gene in the process of interest (in this example, development of the malaria parasites in the mosquito midgut was dramatically increased in the dsRNA-treated mosquitoes).

3. Injection mixture: 400 µg/mL of transformation vector and 150 µg/mL of helper plasmid phsp-pBac in injection buffer (*see* **Section** 2.4, Step 2). Prepare three or more aliquots of 20 µL each. Keep at –20°C until use.

2.3. Injection of dsRNA into Adult Mosquitoes

1. Strains of *A. gambiae* can be obtained from the MR4 (*see* **Note 1**). Mosquitoes are bred at 28°C and 70% humidity, with a day/night period of 12 h/12 h.
2. Waxed paper cartons (e.g., ice-cream or drink containers); filter paper circles of a matching diameter to fit in the bottom of the cartons (#1, Whatman, Kent, UK); fine nylon or cotton netting; tape; small elastic bands.
3. Mosquito aspirator.
4. Absorbent cotton wool; 10% sugar solution in water; small-sized Petri dish bottoms.
5. CO_2 bottle, CO_2 distributor, and pad (InjectMatic, Geneva, Switzerland).
6. Injector Nanoject II (Drummond, Broomall, USA); capillaries (FT330B); a syringe (1 mL) and needle (G24) filled with mineral oil; a paintbrush, forceps.
7. Micropipette puller (P-97, Sutter Instrument Company, Novato, USA).

2.4. Establishment of Transgenic Mosquitoes Expressing Hairpin RNAs

1. Mosquito embryos.
2. Injection buffer (5 mM KCl, 0.5 mM sodium phosphate, pH 6.8), filter paper, a fine paintbrush, glass slides.
3. Isotonic buffer (150 mM NaCl, 4.9 mM KCl, 10.7 mM, Hepes, 2.4 mM $CaCl_2$, pH 7.2) (*see* **Note 2**).
4. Stereoscope MZ6 (Leica, Wetzlar, Germany) with cold light source CLS150, or similar.
5. Reverted microscope Diaphot or similar, with 5x, 10x, and 20x objectives; NT-88NE three-dimensional micromanipulator (Narishige, Tokyo, Japan) or similar; microinjector Eppendorf femtojet (Eppendorf AG, Hamburg, Germany) or similar; sterile femtotip capillaries (Eppendorf AG, Hamburg, Germany); microloaders (Eppendorf AG, Hamburg, Germany).
6. Small Petri dishes (60 mm), 3MM Whatman filter paper (Whatman, Kent, UK), blotting paper.
7. Fluorescence microscope Nikon Eclipse TE200 (Nikon, Melville, USA) or similar; fluorescence filter sets (Texas Red/Cy3.5 and Blue GFP Bandpass).
8. Incubator at 28°C, 70% humidity with a day/night period of 12 h/12 h.

2.5. Analysis of Efficiency of RNAi Silencing

1. Hand-held homogenizer, TRIzol reagent (Invitrogen, Cergy Pontoise, France), DEPC-treated water, sterile

RNase-free labware (tips, microtubes), chloroform, iso-propanol, 75% ethanol, spectrophotometer.

2. SuperScript III reverse transcriptase and reaction buffer (Invitrogen, Cergy Pontoise, France) or similar, dNTPs, random hexamer primers, RNase OUT (Ambion, Applied Biosystems, Foster City, CA, USA), PCR machine.

3. SybrGreen reaction mix, primers specific for the target gene (*see* **Note 3**).

4. Quantitative PCR machine (*see* **Note 4**).

5. Polyacrylamide gel electrophoresis, running buffer, 6x protein loading buffer (350 mM Tris-HCl (pH 6.8); 10.28% SDS; 36% Glycerol; 5% ß-Mercaptoethanol; 0.012% Bromphenol blue), PageRuler^{TM} Prestained Protein Ladder (Fermentas International Inc, Burlington, Canada).

6. Membrane for protein transfer (Amersham Hybond-P, GE Healthcare UK Ltd, Buckinghamshire, England), wet protein transfer unit (Bio-Rad Laboratories, Hercules, USA), transfer buffer.

7. Washing solution: Phosphate-buffered saline (PBS): prepare a 10x stock (130 mM NaCl; 7 mM, Na_2HPO_4; 3 mM NaH_2PO_4) and autoclave. For working solution, dilute one part with nine parts of water. Blocking solution: PBS, 5% nonfat milk powder (Nestlé S.A., Vevey, Switzerland)).

8. Secondary antibody solution: Use the recommended dilution of anti-rabbit, anti-mouse, or anti-rat IgG antibodies conjugated with horseradish peroxidase (HRP).

9. Amersham ECL^{TM} Western Blotting Detection Reagents (GE Healthcare UK Ltd, Buckinghamshire, England).

10. X-ray film (Fuji Photo Film CO., Ltd, Tokyo, Japan), cassette.

3. Methods

In *Anopheles* mosquitoes, RNAi-mediated silencing can be achieved by two different methods: (a) injecting double-stranded RNA (dsRNA) directly into the body cavity of adult mosquitoes *(2)* (**Fig. 5.1A**); and (b) expressing dsRNA in situ from a stably integrated transgene *(4, 5)* (**Fig. 5.1B**). The method of choice often depends on the application. Direct injection of dsRNA permits a fast assessment of the function of the target gene at a selected developmental stage, allowing rapid medium-scale screens. The relatively high number of mosquitoes needed for injections and the transient nature of silencing of some genes are among the most common limitations of this method.

The generation of transgenic lines expressing stable RNAi transgenes is more labor intensive; however, it provides an inexhaustible supply of mutant mosquitoes for in-depth phenotypical and biochemical analyses, and allows the time and tissue-specific knockdown of the target genes through the use of appropriate promoters.

Direct injection of dsRNA includes several steps: (a) a candidate gene is selected; (b) the optimal target sequence is cloned into an appropriate vector for dsRNA synthesis; (c) dsRNA is injected into adult mosquitoes and the efficiency of gene silencing is examined 1–4 days after injection. In general, the whole process, from the selection of a gene of interest to elucidation of its function in the particular biological process, may be achieved in 1 month. Understandably, establishment of transgenic mosquitoes requires longer periods before any functional tests can be performed. This includes the sometimes delicate cloning into the *piggyBac* expression vectors; injections of a large number of embryos; efficient crossing methods to obtain a sufficiently large G_1 progeny; screening of G_1 individuals, normally at the larval stage; identification of positive, transgenic individuals; and the establishment of transgenic lines. Initial functional analysis can be performed with heterozygous lines, provided a screening is carried out to ensure the transgenic origin of the individuals selected for the experiments. In all, it might take 2–3 months from the identification of the gene of interest to the establishment of the transgenic line(s).

3.1. dsRNA Synthesis

1. Clone a fragment of a gene of interest in pLL10 (**Fig. 5.1A**). The optimal size of the fragment is determined by the efficiency of the RNA synthesis reaction, and is between 100 and 1000 bp for the T7 MEGAscript kit. To avoid off-target effects, it is important to design dsRNA constructs for at least two highly specific regions of the gene of interest. Potential gene cross-silencing can be gauged using the DEQOR software (http://cluster-1.mpi-cbg.de/Deqor/deqor.html) *(6)*.
2. Prepare DNA of the constructed plasmid (*see* **Note 5**).
3. For the synthesis of sense and antisense ssRNAs, 2 × 10 μg of plasmid are linearized separately with two different restriction enzymes, one on each side of the insert, in 50 μL (**Fig. 5.1A**). Confirm that digestion is complete by loading 1 μL of the reaction on 1% agarose gel (*see* **Note 6**).
4. Add 25 μL of proteinase K final solution and 4 μL of 10% SDS to each digest. Incubate at 50°C for 30 min.
5. Add 80 μL of phenol/chloroform/isoamyl alcohol. Vortex and incubate 2 min at room temperature (RT). Centrifuge

for 5 min at maximal speed and collect the aqueous phase in a fresh tube.

6. Add 80 μL of chloroform. Vortex and incubate for 2 min at RT. Centrifuge at maximal speed and collect aqueous phase in a fresh tube.

7. Add 56 μL of isopropanol, mix by inverting the tube, and incubate at 4°C for 15 min. Centrifuge for 15 min at 4°C at maximal speed and discard supernatant.

8. Wash with 100 μL of 70% ethanol, centrifuge 5 min at maximal speed, discard supernatant and allow the DNA pellet to air dry on the bench.

9. Dissolve pellet in 20 μL of water. Check DNA quality and concentration (should be around 0.5 μg/μL) on a 1% agarose gel. Linearized plasmids can be stored at −20°C.

10. Thaw reagents of the T7 MEGAscript kit (Ambion, Applied Biosystems, Foster City, CA, USA). Keep ribonucleotides on ice, and transcription buffer at RT.

11. For each plasmid, add in order: $(8-x)$ μL of water; 2 μL of each NTP; 2 μL buffer; x μL linearized plasmid (1 μg), and 2 μL enzyme mix. If several reactions are assembled in parallel, prepare a master mix with all reagents except linearized plasmids, aliquot calculated volume in each tube, then add plasmids. Incubate overnight at 37°C (8–14 h). *See* **Notes** **7** and **8**.

12. DNA template is then digested by adding 1 μL of DNase I to each reaction. Incubate 15 min at 37°C.

13. Purification of ssRNAs: To the incubation mix add 115 μL of water and 15 μL of ammonium acetate stop solution and mix thoroughly. Extract RNA with 150 μL of phenol/chloroform/isoamyl alcohol, followed by 150 μL of chloroform (*see* Steps 5 and 6). Recover aqueous phase and transfer it to a fresh tube.

14. Precipitate RNA by adding 150 μL of isopropanol, mix well by inverting the tube and incubate 15 min at −20°C. Centrifuge 15 min at 4°C at maximal speed and discard supernatant. Air dry pellet for a few minutes on the bench and dissolve it in 20 μL of water.

15. While samples are drying, boil water in a 1–3 L beaker covered with aluminum foil.

16. Measure concentration of ssRNAs using a UV spectrophotometer. Dilute 1 μL of each ssRNA in 9 μL of 10 mM Tris buffer (pH 8). Take 1 μl of this dilution into 100 μl of the same 10 mM Tris buffer for measurement, and keep the remaining 9 μl at −20°C. To calculate the concentration, use the following formula: Concentration of ssRNA in μg/μL $=$ OD$_{260}$ × dilution factor × 40/1000. The dilution factor is 1000 in the example given.

17. Adjust concentration of sense and antisense ssRNAs to 3 μg/μL each and mix equal volumes (the rest of the ssR-NAs can be stored at –20°C or –80°C). Close tubes tightly and boil samples for 5 min in the beaker. Allow samples to slowly cool down to RT.

18. Check dsRNA quality on a 1.8% agarose gel. For this, denature the ssRNA dilutions from Step 16 at 95°C for 3 min and immediately cool on ice. Spin briefly. Run 1 μl of these and 1 μl of dsRNA from Step 17 (in 5 μL of DNA loading buffer) on gel. A clear shift should be observed in the migration patterns of ssRNAs and dsRNA: dsRNA migrates slower than the corresponding ssRNAs.

19. dsRNA is quite resistant to multiple freezing/thawing cycles. Store dsRNA –20°C or –80°C.

3.2. dsRNA Injection into Adult Mosquitoes

1. Mosquito breeding: *See* "Anopheles Culture" by M.Q. Benedict, CDC Atlanta, USA at the MR4 website (http://www2.ncid.cdc.gov/vector/vector.html).

2. Needles for injection: Pull glass capillaries using the needle puller. Each capillary gives two needles in which elongated tip is sealed.

3. Prepare mosquito pots. Cut a cross-shaped opening in the side of a pot and seal it with tape. Place a filter paper circle at the bottom of the pot; it will blot out excess of sugar or mosquito droppings. Stretch netting over the top of the pot and secure with an elastic band.

4. Place around sixty 1–2-day-old female mosquitoes into the pot using an aspirator. Fill a small-sized Petri dish bottom with a cotton pad soaked in 10% sugar solution and place it on the netting.

5. Break the tip of the glass capillary with forceps, so that the tip is rigid enough, but the opening not too large. Fill the capillary with mineral oil using a syringe and a needle. Assemble the Nanoject injector and fill capillary with the dsRNA solution. Verify settings on the control block of the Nanoject and set the desired injection volume and speed. To analyze one gene, we usually inject each mosquito with 69 nL of dsRNA at the highest speed.

6. Immobilize mosquitoes in the pot by CO_2 treatment and align them with dorsal side up on the CO_2 pad. Using the injector and a brush, carefully inject dsRNA solution into the dorsal plate of the mosquito thorax (*see* **Notes 9** and **10**). During injections, limit exposure of mosquitoes to CO_2 using the pedal-controlled distributor (*see* **Note 11**).

7. After all mosquitoes are injected, gently place them back in paper pots using a brush, stretch the netting, fix it with an elastic band, and place a sugar cotton pad on the top. Keep

mosquitoes until further analyses in the humidified 28°C incubator (usually for 4 days). *See* **Note 12**

3.3. Design and Cloning of the RNAi Expression Plasmid

1. Clone 300–800 bp sense and antisense sequences of the target gene in a tail-to-head fashion, under the control of the appropriate promoter, into an intermediate plasmid of choice. Insert the whole RNAi DNA cassette into the single *Asc* I site of transformation vector pBac[3xP3-EGFPafm] or similar (**Fig. 5.1B**). *See* **Note 13**

3.4. Establishment of Transgenic Lines Expressing Hairpin RNAs

1. For standard techniques of mosquito breeding, see above.
2. Blood feed 3–5-days-old female mosquitoes (approximately 100–150 per cage). Starve females for a few hours prior to blood feeding by depriving them of sugar.
3. At 48–72 h post blood feeding, place an oviposition cup (60 mm Petri dish containing Whatman 3MM filter paper soaked in isotonic buffer) into the cage (*see* **Note 14**).
4. Collect embryos every 30 min, each time placing a new oviposition cup into the cage to allow females to lay more embryos. After removing the cup containing the embryos from the cage, place it at RT to slow down development and hence hardening of the chorion.
5. Cut a piece of membrane and a piece of blotting paper with a razor blade at a 135° angle, as shown in **Fig. 5.1B**. Place the membrane (top) and the blotting paper (bottom) onto a glass slide, with the blotting paper slightly exposed, and moisten them with isotonic buffer.
6. At 80–100 min after oviposition, when embryos have started to darken slightly, use a fine paint brush to align 20–40 embryos on the glass slide against the moistened blotting paper, with their posterior poles oriented towards the top of the glass slide, as shown in **Fig. 5.1B**. All embryos must be oriented in the same direction. Remove excess of liquid; however, always keep the membrane moist adding buffer from time to time to avoid embryo desiccation.
7. Transfer the embryos to the microscope for injection. Fill an Eppendorf femtotip glass needle with 2 μL of injection mixture, using microloaders. Make sure the needle is open before injection, by checking for the presence of a little drop at the tip of the needle after injection (*see* **Note 15**).
8. Inject the posterior part of the embryo, using an injection pressure (pi) of 700–1000, and a counter pressure (pc) of 300–500, depending on the needle. When injected successfully, a slight clearing of the embryo yolk should be visible. Take care to remove the embryos that have not been injected, as these are the most likely to survive and therefore their presence will increase the screening efforts (*see* **Note 16**).

9. After injection, place the slide directly onto a tray containing water and larval food, placing strips of 3MM paper around the side of the tray to avoid embryos getting dry. After 2–4 days, count the embryos that have hatched.

10. Separate females from males immediately after emergence or by sexing them at the pupal stage in order to ensure they are virgins (G_0 generation). Cross G_0 adults with wild-type (WT) individuals, placing 10–15 G_0 females with 30–50 WT males and 10–15 G_0 males with 30–50 WT virgin females in separate cages (**Fig. 5.1B**). Blood feed 4–6 days after the crosses have been set up to provide sufficient time for mating, and collect G_1 progeny. In order to provide an estimate of fertility of the G_0 individuals, females can be forced to lay eggs individually in Petri dishes or in 50 mL Falcon tubes in which the conical bottom part has been cut off and replaced by a fine netting (to allow sugar feeding through the use of cotton pads or paper soaked in 5% sucrose solution), while the lid can function as a small oviposition cup when filled with water. Repeat the feedings/egg collections a minimum of three times per cross, especially with the crosses involving G_0 males, in order to increase the number of progeny screened.

11. Screen G_1 progeny at the larval stage with the appropriate fluorescence filter, separate transgenic individuals from negative ones and backcross them to WT individuals of the opposite sex to propagate the line (**Fig. 5.1B**).

12. Sequence some individuals from each transgenic line to ensure the integrity of the RNAi cassette.

3.5. Analysis of Efficiency of RNAi Silencing at the Transcriptional Level

1. Collect 10–15 mosquitoes from the control and experimental groups at different time points after dsRNA injections or at the desired developmental stage in the case of transgenic individuals (*see* **Note 17**), and extract total RNA using TRIzol reagent or equivalent according to the manufacturer's instructions. Measure the concentration of the RNA solution.

2. Convert 2 μg of total RNA into cDNA using SuperScript reverse transcriptase or similar as recommended by the manufacturer (*see* **Note 18**).

3. Perform quantitative polymerase chain reaction (q-PCR) using primers for the gene of interest and SybrGreen PCR master mix. Compare levels of the target gene expression in the control and experimental mosquitoes (**Fig. 5.1C**) (*see* **Note 19**).

3.6. Analysis of Efficiency of RNAi Silencing at the Protein Level

1. Immunoblotting can be performed on hemolymph extracts or on other relevant tissues (**Fig. 5.1C**). For hemolymph collection, clip the proboscis of 10 mosquitoes and collect

clear drops of blood directly into a 6x protein-loading buffer. As an example, for total tissue extracts, grind five mosquitoes in 250 μL of protein extraction buffer, centrifuge the sample at 4°C at 3,000 rpm to clear the extract from remaining tissues, and mix 5 μL of the cleared sample with 5 μL of the 6x protein loading buffer (*see* **Note 20**).

2. Run protein extracts on SDS-PAGE using an appropriate polyacrylamide concentration until the loading buffer dye reaches the bottom of the gel.

3. Transfer proteins from the gel to a membrane according the instructions provided with the protein transfer unit, and incubate the membrane with the primary antibody against the protein of interest.

4. Detect the primary antibody using a secondary HRP-conjugated antibody according to general protocols, and develop the membrane using ECL™ Western Blotting Detection Reagents according to the manufacturer's instructions.

5. Wrap the wet membrane in plastic and expose it on an X-ray film for 30 s. Adjust the exposure time according to the strength of the signal.

4. Notes

1. MR4 (Malaria Research and Reference Reagent Resource) stores and provides reagents to the malaria research community; *see* http://www.malaria.mr4.org/.

2. Solutions are stored at RT, unless otherwise stated.

3. The choice of the reaction mix depends on the q-PCR machine; in our experience, testing a number of primer concentrations greatly improves efficiency.

4. Similar results are obtained with different brands of q-PCR machines.

5. Sequencing of an insert cloned in pLL10 can be done using universal M13 forward and reverse primers.

6. All plasmid DNA should be digested as RNA polymerases are very processive and will generate long heterogeneous transcripts from circular plasmids.

7. When transcription is optimal, the reaction at the end is rather viscous.

8. Sense and antisense ssRNAs can be produced in a single reaction, e.g., from a PCR fragment that was amplified with primers bearing T7 promoters. However this does not ensure that both strands are synthesized with the same efficiency. We therefore prefer to prepare sense and antisense

ssRNAs separately and measure their respective concentrations before annealing, to make sure equal quantities of both strands are mixed and to obtain reproducible quantities of dsRNA.

9. The maximum volume that can be injected at once is 69 nL. To inject larger volumes, repeated injections (up to four) can be performed.

10. With one filled needle, about 60 mosquitoes can be injected with 69 nL each.

11. Excess of CO_2 treatment is toxic to mosquitoes.

12. Humidity improves survival of injected mosquitoes. Make sure to place mosquitoes into a humidified incubator or chamber immediately after injection.

13. To facilitate cloning of the sense and antisense inverted repeats in *E. coli*, it may be necessary to insert a small linker region or an intron between them *(5)*.

14. In the case of *Anopheles stephensi* mosquitoes, it is advisable to soften their chorion before injection to facilitate the injection procedure. To this end, embryos are laid in a 0.1 mM *p*-nitrophenyl *p'*-guanidinobenzoate (pNpGB) (Sigma-Aldrich, St. Louis, USA) solution, dissolved in isotonic buffer, and kept there until injection *(7)*.

15. Many laboratories prefer to pull their own capillaries and use quartz needles for the injection procedure *(8)*.

16. During a set of injections, it is likely that the needle will get clogged due to cytoplasm uptake. It may then be necessary to use the "Clean" function on the Femtojet (reaching high pressure values) to try to unclog the needle, and in some cases the same function may be used for the actual injection procedure.

17. Collected samples can be kept at −80°C for several weeks until further use.

18. To obtain sufficient amounts of cDNA the reaction volume can be scaled up to 100 μL.

19. Not all genes show significant differences at the transcriptional level. In some instances (e.g., *Cactus*) reduction of 20% in the transcriptional level is sufficient to trigger a prominent phenotype *(9)*.

20. Do not store the remaining samples, as protein extracts are not stable and rapidly degrade.

Acknowledgments

The authors first developed these methodologies while working at the EMBL in the group of Professor Fotis C. Kafatos (dsRNA injections) and in the laboratory of Professor Andrea

Crisanti at Imperial College London (transgenesis). They further acknowledge members of the London and Strasbourg laboratories for constructive discussions. This work was supported by grants from CNRS, Inserm, Wellcome Trust, BBSRC, Schlumberger Foundation For Education and Research (FSER), and by the 6th European Commission Programme "Networks of Excellence" BioMalPar. F.C. is a MRC Career Development fellow. E.A.L. is an International Scholar of the Howard Hughes Medical Institute.

References

1. van Dijk, M. R., Waters, A. P., and Janse, C. J. (1995) Stable transfection of malaria parasite blood stages *Science* **268**, 1358–1362.
2. Blandin, S., Moita, L. F., Köcher, T., Wilm, M., Kafatos, F. C., and Levashina, E. A. (2002) Reverse genetics in the mosquito *Anopheles gambiae*: targeted disruption of the Defensin gene *EMBO Rep* **3**, 852–856.
3. Handler, A. M., and Harrell, R. A., 2nd (1999) Germline transformation of *Drosophila melanogaster* with the piggy-Bac transposon vector *Insect Mol Biol* **8**, 449–457.
4. Brown, A. E., Bugeon, L., Crisanti, A., and Catteruccia, F.(2003) Stable and heritable gene silencing in the malaria vector *Anopheles stephensi Nucleic Acids Res* **31**, e85.
5. Brown, A. E., Crisanti, A., and Catteruccia, F.(2003) Comparative analysis of DNA vectors at mediating RNAi in Anopheles mosquito cells and larvae *J Exp Biol* **206**, 1817–1823.
6. Henschel, A., Buchholz, F., and Habermann, B.(2004) DEQOR: a web-based tool for the design and quality control of siRNAs *Nucleic Acids Res* **32**, W113–W120.
7. Catteruccia, F., Nolan, T., Loukeris, T. G., Blass, C., Savakis, C., Kafatos, F. C., and Crisanti, A.(2000) Stable germline transformation of the malaria mosquito *Anopheles stephensi Nature* **405**, 959–962.
8. Lobo, N. F., Clayton, J. R., Fraser, M. J., Kafatos, F. C., and Collins, F. H.(2006) High efficiency germ-line transformation of mosquitoes *Nat Protoc* **1**, 1312–1317.
9. Frolet, C., Thoma, M., Blandin, S., Hoffmann, J. A., and Levashina, E. A. (2006) Boosting NF-kappaB-dependent basal immunity of *Anopheles gambiae* aborts development of *Plasmodium berghei*. *Immunity* **25**, 677–685.

Chapter 6

RNAi Using a Chitosan/siRNA Nanoparticle System: In Vitro and In Vivo Applications

Morten Østergaard Andersen, Kenneth Alan Howard,, and Jørgen Kjems

Abstract

Delivery is a key issue in development of clinically relevant RNAi therapeutics. Polymeric nanoparticles formed by self-assembly of polycations with siRNA can be used for extracellular delivery, cellular uptake and intracellular trafficking as a strategy to improve the therapeutic potential of siRNA. This chapter describes a chitosan-based nanoparticle system for in vitro and in vivo transfection of siRNA into cells. The method exploits the mucoadhesive and mucopermeable properties of this cationic polysaccharide to deliver siRNA across mucosal epithelium and provides a platform for targeting human diseases with RNAi therapeutics.

Key words: siRNA, Chitosan, Nanoparticles, Macrophages, Nasal Delivery, Intraperitoneal Delivery, TNFα, Freeze Drying.

1. Introduction

RNA interference (RNAi) has proven to be an effective method to knockdown expression of individual genes by the action of double-stranded RNA *(1, 2)* adding to renewed optimism that it will be possible in the foreseeable future to use gene inhibitory medicine to treat human disease. The most prominent obstacle is delivery of the large siRNA molecule into diseased cells of the patient. Polycationic polymer-based nanoparticle (or polyplex) systems used for site-specific delivery, cellular uptake and intracellular trafficking of plasmid DNA *(3, 4)* can be adopted to improve the therapeutic potential of siRNA. This chapter describes a chitosan-based nanoparticle system for in vitro and in vivo

John F. Reidhaar-Olson and Cristina M. Rondinone (eds.), *Therapeutic Applications of RNAi: Methods and Protocols, vol. 555*
© Humana Press, a part of Springer Science+Business Media, LLC 2009
DOI 10.1007/978-1-60327-295-7_6 Springerprotocols.com

silencing applications *(5, 6)*. The mucoadhesive *(7)* and mucoper-meable *(8)* properties of this cationic polysaccharide are exploited for siRNA delivery across mucosal epithelium as an approach to target diseases of the mucosa and overcome serum-induced breakdown and first-pass clearance associated with intravenous administration of nanoparticles. Furthermore, chitosan-mediated transfection by a mannose-like lectin receptor *(9)* interaction facilitates delivery into macrophages and silencing of cytokines such as tumour necrosis factor alpha (TNFα) that are involved in inflammatory and infectious diseases. Intraperitoneal administration into a serum-free macrophage-rich environment can be used as a strategy to utilize this property for in vivo applications *(10)*.

2. Materials

2.1. Nanoparticle Preparation, Storage and Drying

1. 60% sucrose solution.
2. Filter (0.2 μm) (Sartorius, Goettingen, Germany).
3. siRNA, including fluorescently labelled siRNA (Dharmacon, Co, USA, DNA Technology, Aarhus, Denmark and Integrated DNA Technologies (IDT), IA, USA).
4. Solubilization buffer: 0.2 M NaAc, pH 4.3. Filter through a 0.2-μm filter, make with nuclease-free water.
5. Nanoparticle Formation Buffer: 0.2 M NaAc, pH 5.5 pH adjusted with 1 M NaOH, make with nuclease-free water.
6. Chitosan Solution: <5 mg/ml Chitosan (Bioneer, Hørsholm, Denmark or Novomatrix, Sandvika, Norway) in 0.2 M NaAc, pH 5.5. Dissolve chitosan in solubilization buffer overnight, filter through a 0.2-μm filter and adjust pH to 5.5 with 1 M NaOH (*see* **Notes 1** and **2**).

2.2. Transfection and Reverse Transfection

1. Complete media: RPMI Medium (RPMI) supplemented with 10% fetal bovine serum and 1% penicillin/streptomycin. Add 0.5% G418 if cells are expressing a plasmid with a neomycin resistance gene.
2. Phosphate-buffered saline (PBS): Prepare 10X stock with 1.37 M NaCl, 27 mM KCl, 100 mM Na_2HPO_4, 18 mM KH_2PO_4. Adjust to pH 7.4 with 1 M HCl and autoclave before storage at room temperature.

2.3. Flow Cytometry

1. PBS (prepared as described above).
2. Trypsin-EDTA.
3. 1% Paraformaldehyde in PBS.

2.4. Cytotoxicity

1. CellTiter 96® AQueous One Solution Cell Proliferation Assay (Promega, Wisconsin, USA).

2.5. Tumour Necrosis Factor Alpha (TNFα) ELISA

1. Goat anti-mouse TNFα capture antibody, recombinant murine TNFα, goat anti-mouse TNFα detection antibody, streptavidin-horseradish peroxidase, tetramethylbenzidine (R&D Systems).
2. Lipopolysaccharide (LPS) (InvivoGen, Ca, USA).
3. Bovine serum albumin (BSA)
4. Coating buffer: 15 mM Na$_2$CO$_3$, 35 mM NaHCO$_3$, 0.02% NaN$_3$ (pH 9.6).
5. Blocking buffer: PBS with 5% sucrose and 0.05% NaN$_3$ (pH 7.4).

2.6. Extraction and Isolation of Primary Macrophages

1. MEM Media.

2.7. Pulmonary Gene Silencing by Nasal Administration

1. Isoflurane.
2. Zoletilmix/Torbugesic mix: 0.5 ml zoletilmix, 1.3 ml sterile water, 0.7 ml torbugesic (diluted prior 1:100 in sterile water), dosage is ∼0.14 ml for an animal weighing 22 g.
3. DAPI (Sigma, St. Louis, MO, USA).
4. VivaSpin20 centrifugal concentrators (MW cut-off 100 kDa) (Vivascience, USA)

2.8. Intraperitoneal Transfection of Peritoneal Macrophages

1. Syringes (1 ml) (Terumo, NJ, USA).
2. Needles 25G, 0.5 × 25 mm (BD, NJ, USA).

3. Methods

3.1. Nanoparticle Preparation

The formation of interpolyelectrolyte complexes (nanoparticles or polyplexes) occurs between anionic siRNA duplexes and cationic chitosan polymer *(5, 6)*.

1. Dilute chitosan solution to 0.8 mg/ml with particle formation buffer.
2. Stir 1 ml in a tube (internal diameter = 13 mm, internal length = 96 mm) containing a small magnetic fly (length = 7.5 mm, width/height = 1.5 mm).
3. Add 20 μl of 100 μM siRNA in distilled water slowly in one slow continual action from a pipette tip that has been inserted into the liquid while stirring at medium speed on a bench stirrer (*see* **Note 3**). Using the aforementioned amounts of chitosan (84%DD) and siRNA will result in a N:P ratio of ∼63 (*see* **Note 4**).
4. Leave the solution to stir for an hour after which it is ready for use.

3.2. Storage and Drying

After nanoparticle formation it is possible to store the particle at 4°C for 1 week. But in that case, the particle solution should be mixed gently prior to use.

If an aqueous lyoprotectant such as sucrose is added to the nanoparticles at a final concentration of 5% or more (e.g. by adding 200 µl 60% sucrose to a 1 ml particle solution) the particles can be frozen and thawed later without loss of activity; they can also be vacuum dried and freeze dried (11). If a lyoprotectant is not added the nanoparticles cannot be dried nor frozen without a complete loss of activity. If freeze drying is desired the particles are frozen to –20°C and freeze dried at –20°C and under 200 mT (Duradry/Durastop freezer dryer system, FTS Systems, NY, USA). The nanoparticles can be vacuum dried without freezing but a small loss (10–20%) in transfection efficiency will then be observed. After drying the nanoparticle/sucrose solution it should become transparent, flat and highly viscous. After drying, the nanoparticles are best stored dry, in the dark at 4°C or at –20°C for optimal storage, but the nanoparticles are still capable of 40% knockdown after 2 weeks, if stored at 25°C.

3.3. Transfection

1. Seed the cells the day before transfection at a concentration of 25,000–55,000 cells/cm^2 in complete media (500 µl for 24-well plates, 100 µl for 96-well plates).
2. On the day of transfection remove the media and add fresh complete media. The volume of complete media added should be 250 µl for 24-well plates and 100 µl for 96-well plates minus the volume of transfection agent added.
3. Add the transfection agent to give a 50 nM siRNA concentration in the wells (for batch of nanoparticles made with 20 µl 100 µM siRNA in 1 ml 0.8 mg/ml chitosan one should add 6.4 µl nanoparticle solution to 244 µl media) (see **Note 5**).
4. Remove the transfection media from the cells after at least 4 h and within 24 h (see **Note 6**).
5. Wash the wells with PBS (500 µl or more for 24-well plates, 100 µl or more for 96-well plates).
6. Replace the PBS with fresh complete media (500 µl or more for 24-well plates, 100 µl or more for 96-well plates).

3.4. Reverse Transfection

It is an efficient gene silencing method with potential applications for high throughput gene screening tool and longer shelf-life use (11).

1. Add complete media containing 25,000–100,000 cells/cm^2 to the wells. The amount of nanoparticles dried in each well and the volume of media added should be set so the final concentration in each well equals 50 nM (e.g. adding 250 µl media with 100,000 cells to each well on

a freeze dried 24-well plate where in each well was added 7.6 μl chitosan/siRNA/sucrose solution (composed of 1 ml 0.8 mg/ml chitosan, 20 μl 100 μM siRNA and 200 μl 60% sucrose)).

2. Mix the wells gently.

3. After 24 h remove the media and replace it with new growth media (500 μl or more for 24-well plates, 100 μl or more for 96-well plates) (*see* **Note 7**).

3.5. Flow Cytometry

To determine and quantify the transfection and knockdown efficiency in a cell line flow cytometry can be performed; this either requires that the cell line expresses a fluorescent protein (such as enhanced green fluorescent protein (eGFP) endogenously, or that the cell line has been transfected with a plasmid expressing such a protein prior to or at the same time as the knockdown study (known as a co-transfection, in which case a plasmid transfection agent is also needed). Alternatively the cells can be transfected with siRNAs that have been fluorescently labelled with dyes such as Cy3 or Cy5 (Dharmacon, Co, USA). In this case flow cytometry measures the cellular binding and uptake of the siRNA molecules.

To perform the flow cytometry the transfected cells are typically harvested after 48 or 72 h, suspension cells can be harvested directly (jump to Step 5), adherent cells are harvested using the following protocol.

1. Remove media.

2. Wash each well with 500 μl PBS (24-well plates) or 100 μl PBS (96-well plates).

3. Add 250 μl trypsin-EDTA and incubate for ~5 min at 37°C.

4. Remove trypsin-EDTA and add of 500 μl complete media (for 24-well plates) or 100 μl complete media (for 96-well plates).

5. Transfer the media to 1.5-ml tubes and centrifuge for 5 min at 1500 rpm.

6. Remove media and wash the cells with 500 μl PBS.

7. Centrifuge for 5 min at 1500 rpm.

8. Remove PBS and add 500 μl PBS with 1% paraformaldehyde or formaldehyde.

The cells are then stored at 4°C in the dark between 1 and 7 days before flow cytometry is performed. Flow cytometry is carried out by first selecting the cell population in a dot plot of side scatter vs. forward scatter and then measuring the geometric mean of the fluorescence in a histogram of this cell population. The transfection efficiency of the chitosan/siRNA nanoparticles can then be calculated as 1 minus the geographical mean of match-transfected cells divided by the geographical mean of mismatch-transfected cells (*see* **Note 8**).

3.6. Cytotoxicity

To investigate whether the chitosan/siRNA nanoparticles induce toxicity in a specific cell line, a cytotoxicity assay can be carried out.

1. After 48 h of transfection, remove media and add 100 μl complete media (96-well plate).
2. Add 20 μl CellTiter 96® AQueous One Solution Cell Proliferation Assay to wells containing cells and a "no cells" control.
3. Incubated the cell for 1.5 h at 37°C and 5% CO_2 (*see* **Note 9**).
4. Read absorbance at 490 nm and 600 nm (*see* **Note 10**).

The $\text{absorbance}_{490\text{ nm}} - \text{absorbance}_{600\text{ nm}}$ is directly correlated to the viability of the cells, and the toxicity of the particles can be inferred by comparing this value for transfected cells and non-transfected control cells.

3.7. TNFα ELISA

The determination of TNFα secretion can be used as a measure of downregulation of inflammatory responses in primary macrophages.

1. Harvest the cells after transfection is ended as if for flow cytometry.
2. Add 100 μl of the cell suspension to each well on a 96-well plate for ELISA and another 100 μl cell suspension to a 96-well plate for a cytotoxicity assay.
3. Add 10 μl of 1 mg/ml LPS to the wells and incubate the plate at 37°C and 5% CO_2 for 5 h.
4. Meanwhile coat maxisorp plates overnight at 25°C with 100 μl of goat anti-mouse TNFα capture antibody in a concentration of 2 mg/ml in coating buffer.
5. Block the wells for 1 h with 300 μl of 1% BSA in a blocking buffer.
6. Add 100 μl of supernatants from the cells incubated with LPS to the maxisorp.
7. Add control recombinant murine TNFα to the coated maxisorp plate wells (100 μl in each well).
8. Incubated overnight at 4°C.
9. Subsequently, incubate the wells 25°C for 2 h with 100 μl biotinylated goat anti-mouse TNFα detection antibody (150 ng/ml).
10. Dilute and add streptavidin-horseradish peroxidase 1:200 in TBS with 0.1% BSA.
11. Incubate the mixture for 20 min.
12. Add H_2O_2 and tetramethylbenzidine for colour development.
13. Incubate the plates in the dark for 10–20 min.
14. Stop the reaction by addition of 50 μl 5% H_2SO_4.
15. Read the absorbance at 450 and 570 nm; the latter wavelength is used as a reference. Between each step the plates

are washed three times with PBS-0.05% Tween 20, pH 7.4. As mentioned earlier the measured quantities of TNFα should be normalized to the viability of each sample.

3.8. Extraction and Isolation of Primary Macrophages

For evaluation of particle transfection and gene silencing in "*difficult-to-silence*" primary cells:

1. Kill adult mice by cervical dislocation.
2. Inject 5 ml of MEM containing 20% FBS intraperitoneally with a 25G needle.
3. Agitate the abdomen gently while the peritoneum is exposed and breached.
4. Remove the medium using a syringe.
5. Centrifuge the medium for 10 min at 2500 rpm.
6. Resuspend the pellet resuspended in MEM containing 50% FBS.
7. Plate the suspension on a multiwell 12-well plate.
8. Allow the macrophages to adhere for 2 h before medium (containing non-adherent cells) is removed.
9. Add fresh medium containing 5% penicillin/streptomycin to the cells.
10. After 40 h remove the medium: the macrophages are now isolated and ready for transfection (same particle system 20 μl 100 μm siRNA described earlier).

3.9. Pulmonary Gene Silencing by Nasal Administration

The route of administration is an important determinant for successful RNA-based silencing therapeutics. Local administration to the mucosal surfaces lining the respiratory tract is an attractive alternative to the intravenous route for the treatment of pulmonary diseases.

1. Anaesthetize the mice with isoflurane.
2. Place the mice on their back and administer 30 μl particles (made as above but with 250 μM siRNA, concentrate to 1 mg/ml siRNA using VivaSpin centrifugal concentrator) intranasally (15 μl in each nostril) each day for 5 days (*see* **Note 11**).
3. On day 6 the mice are anaesthetized with an injection of 0.14 ml zoletilmix/torbugesic mix.
4. Perfusion fixate manually with 4% formaldehyde phosphate-buffered solution.
5. Harvest lungs.
6. Paraffin-embed the lungs.
7. Cut exhaustively in 3-μm sections.
8. Sample every 100th section together with the next.
9. Transfer sections into DAPI solution for counterstaining.
10. Wet mount on a Super Frost slide.
11. Analyse the slides in a fluorescence microscope with a UV/GFP filter, a 20x objective, a mounted digital

camera and a motorized stage in conjunction with CAST software.

12. Count the number of EGFP-expressing epithelial bronchial cells by a physical fractionator.

3.10. Intraperitoneal Transfection and TNFα knockdown in Peritoneal Macrophages

Investigation of transfection and gene silencing in systemic macrophages by the intraperitoneal injection. The intraperitoneal route facilitates uptake of particles into systemic macrophages in a serum-free environment.

1. Inject 200 μl of particles containing fluorescently labelled siRNA (N:P ∼ 63 prepared as above) into the peritoneal cavity of mice with a 25G needle.

2. Massage the abdominal area to facilitate the distribution of the particles within the peritoneum.

3. Peritoneal macrophages are isolated, according to the procedure described above, 2 h post-injection and plated in 96-well plates or mounted with DAPI nuclear stain onto glass-bottom dishes (MatTek Corp. Ma, USA) or glass coverslips in 12-well plates.

4. Uptake of fluorescently labelled siRNA is monitored by a Zeiss semi-confocal epifluorescence microscope (or equivalent).

5. TNFα levels in the harvested supernatants of LPS-stimulated macrophages are measured 24 h post-injection according to the above protocol.

---

4. Notes

1. If a higher concentration of chitosan or a buffer with a higher pH is used when dissolving the chitosan it has a tendency to dissolve incompletely.

2. We suggest using a chitosan with a molecular weight over 100 kDa and with a degree of deacetylation (DD) of at least 80%. DD refers to percentage of free amino groups (cationic charges) on the polymer chain after the process of deacetylation of chitin.

3. It is very important that the whole volume of the liquid is being stirred. The volume of the reagents can also be scaled down to 500 μl or 250 μl; in this case, it is, however, even more important to ensure that the liquid is being stirred during the entire process and that continual addition of siRNA is not compromised by hitting the flea with the pipette tip as this can affect particle formation.

4. The N:P ratio refers to the ratio of positive charges on the chitosan amines to the negative charges on the phosphates in the siRNA backbone. To calculate N:P ratios

a mass-per-charge of 325 Da was used for RNA and mass-per-charge of 168 Da was used for chitosan (at 84% deacetylation). The concentration of the chitosan and siRNA can be scaled down and the N:P ratio changed slightly to 1 ml 0.2 mg/ml chitosan and 20 μl 20 μM siRNA; however, particles produced by the aforementioned concentrations show high transfection and silencing efficiency.

5. The complete media can be mixed with the particles prior to addition to the wells rather than adding the media first and then the particles separately to each well.

6. The transfection efficiency and toxicity of the particles appear to vary between different cell types; we suggest limiting the transfection time to 4 h if toxicity is observed.

7. The 24-h transfection period is necessary as the cells need time to attach before the media is changed.

8. If flow cytometry is to be carried out on the same day as cell harvesting, the fixation agent should not be included. Instead it is a good idea to include 2% serum in the PBS to stabilize the cells until and during the flow cytometry. Store the fixed cells in darkness at 4°C. It is always best to include two additional controls containing non-transfected fluorescent and non-fluorescent cells. The geographical mean of the non-fluorescent cells should then be subtracted from the geographical means of the other samples before further calculations are carried out. Non-specific knockdown induced by the particles can be inferred by comparing the geographical mean of the mismatch-transfected cells with that of the non-transfected control cells. If using fluorescently labelled siRNA it is recommended to confirm that the siRNA is taken up by the cells using fluorescent microscopy.

9. Increasing the incubation time will increase absorption time and may be necessary if few cell are present.

10. Other wavelengths can be used but will reduce the sensitivity.

11. Higher N:P with more excess chitosan at this concentration can be used to improve mucoadhesion and mucopermeation in vivo properties but should be tested in an epithelial cell line as the N:P ratio has been shown to affect silencing efficiency (3).

References

1. Fire, A., et al. (1998). Potent and specific genetic interference by double-stranded RNA in *Caenorhabditis elegans*. *Nature* **391**, 806–811.

2. Elbashir, S. M., et al. (2001). Duplexes of 21-nucleotide RNAs mediate RNA interference in cultured mammalian cells. *Nature* **411**, 494–498.

3. Pouton, C. W. and Seymour, L. W. (2001) Key issues in non-viral gene delivery. *Adv. Drug Deliv. Rev.* **46**, 187–203.

4. Elouahabi, A. and Ruysschaert, J. M. (2005) Formation and intracellular trafficking of lipoplexes and polyplexes. *Mol. Ther.* **11**, 336–347.

5. Howard, K. A., et al. (2006). RNA Interference in vitro and in vivo using a novel chitosan/siRNA nanoparticle system. *Mol. Ther.* **14**, 476–484.

6. Liu, X., et al. (2007). The influence of polymeric properties on chitosan/siRNA nanoparticle formulation and gene silencing. *Biomaterials* **28**, 1280–1288.

7. Soane, R. J., et al. (1999). Evaluation of the clearance characteristics of bioadhesive systems in humans. *Int. J. Pharm.* **178**, 55–65.

8. Artursson, P., Lindmark, T., Davis, S. S. and Illum, L. (1994). Effect of chitosan on the permeability of monolayers of intestinal epithelial cells (Caco-2). *Pharm. Res.* **11**, 1358–1361.

9. Feng, J., Zhao, L. and Yu, Q. (2004). Receptor-mediated stimulatory effect of oligochitosan in macrophages. *Biochem. Biophys. Res. Commun.* **317**, 414–420.

10. Howard, K. A., et al. (2009). Knockdown in peritoneal macrophages as an anti-inflammatory treatement in a murine arthritic model. *Mol. Ther.* **17**, 162–168.

11. Andersen, M. O., et al. (2008). Delivery of siRNA from lyophilized polymeric surfaces. *Biomaterials* **29**, 506–512.

Chapter 7

Lentiviral and Adeno-Associated Vector-Based Therapy for Motor Neuron Disease Through RNAi

Chris Towne and Patrick Aebischer

Abstract

RNAi holds promise for neurodegenerative disorders caused by gain-of-function mutations. We and others have demonstrated proof-of-principle for viral-mediated RNAi in a mouse model of motor neuron disease. Lentivirus and adeno-associated virus have been used to knockdown levels of mutated superoxide dismutase 1 (SOD1) in the G93A SOD1 mouse model of familial amyotrophic lateral sclerosis (fALS) to result in beneficial therapeutic outcomes. This chapter describes the design, production, and titration of lentivirus and adeno-associated virus capable of mediating SOD1 knockdown in vivo. The delivery of the virus to the spinal cord directly, through intraspinal injection, or indirectly, through intramuscular injection, is also described, as well as the methods pertaining to the analysis of spinal cord transduction, SOD1 silencing, and determination of motor neuron protection.

Key words: Amyotrophic lateral sclerosis, superoxide dismutase (SOD1), lentivirus, adeno-associated virus (AAV), RNA interference (RNAi), gene therapy, G93A SOD1.

1. Introduction

1.1. Amyotrophic Lateral Sclerosis

Amyotrophic lateral sclerosis (ALS) is a devastating neurodegenerative disorder that results from the progressive and irreversible degeneration of motor neurons in the spinal cord, brain stem, and cerebral cortex. This loss of motor neurons leads to gradual muscle weakness and atrophy, with death often occurring 2–5 years after diagnosis as a result of respiratory failure. Ten percent of ALS cases are familial with one-fifth of those being caused by mutations in the gene encoding for superoxide dismutase 1 (SOD1). Current evidence supports that disease-linked

John F. Reidhaar-Olson and Cristina M. Rondinone (eds.), *Therapeutic Applications of RNAi: Methods and Protocols, vol. 555*
© Humana Press, a part of Springer Science+Business Media, LLC 2009
DOI 10.1007/978-1-60327-295-7_7 Springerprotocols.com

mutations confer SOD1 a toxic gain-of-function, independent of its enzymatic activity, which triggers motor neuron death through a still unclear mechanism *(1)*. In light of this direct toxic effect, reduction of mutated SOD1 protein levels via RNA interference (RNAi) is an attractive therapeutic approach as it targets the cause of the disease regardless of the specific pathogenic mechanism responsible for motor neuron death.

1.2. Viral-Mediated RNAi Against SOD1

RNAi is an innate gene silencing mechanism that utilizes small single- or double-stranded RNA (dsRNA) to promote the degradation of specific mRNA targets, effectively reducing the levels of the corresponding protein *(2)*. Saito et al. *(3)* have demonstrated the potential for RNAi as a therapeutic approach for familial amyotrophic lateral sclerosis (fALS). By crossing transgenic mice expressing dsRNA against SOD1 with the G93A SOD1 mouse model, the development of the ALS-like phenotype was completely prevented with the double transgenic mice having normal spinal cord histology and behaving as wild-type controls.

The current challenge facing RNAi-based therapy for fALS is the delivery of the dsRNA silencing instructions in vivo. Two strategies exist towards this goal. The first approach is to synthesize dsRNA in vitro prior to their in vivo delivery. An example of this is to deliver the dsRNA or antisense oligonucleotides directly to the cerebrospinal fluid (CSF) through ventricular infusion. Indeed, osmotic pumps have been used to deliver antisense oligonucleotides against SOD1 into the ventricles of G93A SOD1 rats leading to silencing of SOD1 within the spinal cord and subsequent therapeutic gains *(4)*. The limiting factor of this intervention, however, is the continual synthesis and delivery of oligonucleotides that would be required to maintain this therapeutic silencing throughout the lifetime of the patient. The second approach, that overcomes this hurdle, is to provide cells with the DNA templates that they may use to indefinitely transcribe short hairpin RNAs (shRNAs) *(5)*. shRNA consist of two complementary RNA strands linked by a hairpin loop that are processed into dsRNA capable of mediating RNAi. Viral vectors can be used to deliver these transgenic cassettes directly to the genome of cells to result in stable and long-lasting expression of shRNA following a single intervention. Three landmark studies have used recombinant viral vectors to deliver shRNAs to result in SOD1 knockdown and achieve therapeutic gain in the G93A mouse model of ALS.

1.2.1. Intramuscular Delivery of Lentivirus Based on EIAV into Newborn SOD1 Mice

Ralph et al. *(6)* delivered lentivirus based on the equine infectious anemia virus (EIAV) to multiple muscle groups of newborn G93A SOD1 mice. As EIAV is capable of efficient retrograde transport from the muscle to the motor neuron soma, this technique facilitated transduction and SOD1 silencing within motor neurons at multiple levels of the spinal cord and brain stem. This resulted in delaying disease onset by 115% (and subsequently increased survival age by 77%) although did not slow disease progression from time of onset.

1.2.2. Intraspinal Delivery of VSV-G-Pseudotyped Lentivirus into Presymptomatic SOD1 Mice

Lentivirus pseudotyped with vesicular stomatitis virus G (VSV-G) were injected directly to the L3 and L4 lumbar spinal cord of presymptomatic 42-day-old G93A SOD1 mice *(7)*. This silenced SOD1 in the motor neurons and glial cells of the ventral horn resulting in a delay of both disease onset and progression, as measured by various electromyographical and behavioral outcomes, in the hindlimb muscles innervated by this region of spinal cord. The observation that this RNAi strategy was capable of retarding disease progression concurs with current evidences that SOD1 expression within glial cells is an important determinant of disease progression *(8)*.

1.2.3. Intramuscular Delivery of AAV Serotype 2 into Presymptomatic SOD1 Mice

Miller et al. *(9)* also capitalized on viral retrograde transport to transduce motor neurons following intramuscular injection, albeit using adeno-associated virus (AAV) serotype 2. Presymptomatic 45-day-old G93A SOD1 mice were injected directly into the lower hindlimb muscles with the vector to result in transduction and silencing of SOD1 in lumbar-level motor neurons. This conferred behavioral improvement in the injected limbs as determined by preservation of grip strength.

The present chapter will detail the experimental procedures required to recapitulate the studies mentioned above. We will describe the design, production, and titration of both VSV-G-pseudotyped lentivirus used in the study by Raoul et al. *(7)* and AAV serotype 6 (AAV6), a vector that we have found to be capable of particularly efficient muscle-to-motor neuron retrograde transport. Next, the intraspinal and intramuscular delivery of these vectors to adult mice will be described, as well as the intramuscular injection to neonates. Finally, the methods required to analyze in vivo transduction, SOD1 silencing, and conferred neuroprotection will be provided.

2. Materials

2.1. Virus Production

1. pSIN-W-PGK: Lentivirus genomic plasmid.
2. pCMVΔR8.92: Packaging plasmid encoding viral genes in *trans*.
3. pMD.G: Plasmid encoding the VSV-G envelope.
4. pRSV-Rev: Plasmid encoding the rev protein of HIV-1.
5. pAAV-CMV-MCS: AAV genomic plasmid.
6. pDF6: Packaging plasmid encoding for the AAV6 capsid.

7. AAV-293 cells (#240073, Stratagene, La Jolla, CA) and 293T cells.

8. Six-well Multiwell™ Tissue Culture Plate ((#166508, Becton Dickinson, Frankin, NJ).

9. 500 cm^2 Nunclon™Δ polystyrene dish (#166508, Nalg Nunc International, Rochester, NY).

10. Cell culture medium: Dulbecco's Modified Eagle's Medium (DMEM, Gibco BRL, Bethesda, MD) supplemented with 10% fetal bovine serum (FBS, HyClone, Odgen, UT). Episerf (Gibco BRL).

11. Trypsinization medium: Hanks' balanced salt solution (HBSS; red phenol 0.4 g/L KCl, 0.06 g/L KH$_2$PO$_4$, 8 g/L NaCl, 0.35 g/L NaHCO$_3$, 0.048 g/L Na$_2$HPO$_4$, 1 g/L D-glucose) containing 0.05% trypsin and 0.53 mM EDTA.

12. Transfection reagents: 0.5 M CaCl$_2$ (73.5 g CaCl$_2$·2H$_2$O; complete to 500 mL with H$_2$O; store in aliquots at –20°C) and 2x HBS (28 mL of 5 M NaCl, 11.9 g of HEPES, 750 μL of 1 M Na$_2$HPO$_4$ in 400 mL H$_2$O; adjust the pH to 7.1 with NaOH and complete to 500 mL; store in aliquots at –20°C).

13. Ultracentrifugation of lentivirus: ultracentrifuge, SW28 rotor, SW60 rotor, and polyallomer tubes (SW28 rotor, 25 × 89 mm, SW60 rotor, 11 × 60 mm, Beckman Coulter, Fullerton, CA).

14. Centricon® Plus-20 centrifuge filters (#UFC2BHK08, Millipore, Billerica, MA).

15. PBS/1% BSA: Add 5 g of bovine serum albumin (BSA, Sigma-Aldrich, St Louis, MO) to 500 mL pH 7.4 phosphate-buffered saline (PBS) and pass through a 0.22-μm filter.

16. 1.6 mL Maxymum Recovery, Clear Microtubes (Axygen Scientific, Union City, CA).

17. 4% paraformaldehyde (PFA): Add 40 g of paraformaldehyde to 900 mL of PBS and heat at 60°C (*see* **Note 1**). Add a few drops of 5 M NaOH to dissolve the paraformaldehyde and allow to equilibrate to room temperature. Filter the solution, adjust pH to 7.4, and complete to 1 L.

18. 5 mL Polystyrene Round-Bottom Tube with Cell-Strainer Cap (#352235, Becton Dickinson).

2.2. Animal Experiments

1. Anesthesia: Anesthesia system, O$_2$ cylinder, induction chamber, surgical table with anesthesia mask, isoflurane.

2. Intraspinal injections: Small Animal Stereotaxic Instrument (Model 963 Ultra Precise, David Kopf Instruments, Tujunga, CA), Mouse Spinal Adaptor (#51690, Stoelting, Wood Dale, IL), Mouse Transverse Clamps (#51695, Stoelting), Scalpel Handle No. 4 (#02-036-040, allgaier instrumente GmbH, Tuttlingen, Germany), Scalpel Blades

(#02-040-021, allgaier instrumente GmbH), Adson Tissue Forceps (#08-265-120, allgaier instrumente GmbH), Derf Needle Holder (#20-116-125, allgaier instrumente GmbH), Bulldog clamps (#18050-28, Fine Science Tools GmbH, Heidelberg, Germany), Friedman-Pearson Rongeur (#16021-14, Fine Science Tools GmbH), self-regulating heating pad (#21061-00, Fine Science Tools GmbH), 27G needle (Terumo®, Leuven, Belgium), 70 mm × 34G (0.23×0.127) needle (Phymed, Paris, France), 50 mm × 26G (0.457×0.254) needle (Phymed), Viscotears® Ophthalmic Gel (Novartis, Basel, Switzerland), Microinjection pump (CMA/100, CMA Microdialysis, Solna, Sweden), Polyethylene Tubing (0.38× 1.1 mm) (A-M Systems, Carlsborg, WA), 6-0 Vicryl™ Suture (Ethicon, Norderstedt, Germany), Betadine® standard solution (Mundipharma GmbH, Basel, Switzerland), electric shaver and 500 mg Paracetamol effervescent tablets.

3. Intramuscular injections of adults: Hamilton 25 μL syringe (#702, Hamilton, Reno, NV), Scalpel Handle No. 4, Scalpel Blades, 6-0 Vicryl™ Suture, Derf Needle Holder, Adson Tissue Forceps, Betadine® standard solution, and electric shaver.

4. Intramuscular injections of neonates: 30G × 1 mm Insulin Needle (Becton Dickinson), Polyethylene Tubing (0.28×0.61 mm) (A-M Systems), and a Hamilton 10 μL syringe (#701, Hamilton).

2.3. Histology

1. Heparin (Liquemin, Roche, Basel, Switzerland).
2. Spinal cord dissection: McPherson-Vannas spring scissors (#52130-01, Stoelting), Friedman-Pearson Rongeur, Tissue Forceps, and blunt-dissection probe.
3. 25% sucrose: Dissolve 25 g of sucrose in 100 mL of PBS and pass through a 0.22-μm filter.
4. Cryosections: Peel-A-Way® Disposable Embedding Molds (#18646A, Polysciences Inc, Warrington, PA), Shandon Cryomatrix™ (#6769006, Thermo Electron Corporation, Pittsburgh, PA), goal polyclonal anti-GFP Ab (ABCAM, Cambridge, UK), rabbit polyclonal anti-GFP Ab (ABCAM), monoclonal anti-SOD1 Ab (clone SD-G6, Sigma-Aldrich), monoclonal anti-SOD1 Ab (clone G215-1, BD Pharmingen, Franklin, NJ), polyclonal goat anti-SOD1 Ab (sc-8636, Santa Cruz, Santa Cruz, CA), rabbit polyclonal anti-VAChT Ab (Sigma-Aldrich).
5. Tris-buffered saline (TBS): Dissolve 12.11 g of Tris-base (MW= 121.14) and 8.76 g NaCl in 1 L of H_2O adjusting the pH to 7.6.
6. Paraffin sections: Cresyl Violet (Sigma-Aldrich), Toluidine Blue (Sigma-Aldrich), osmium tetroxide (Sigma-Aldrich), Leica Histomold (2×15 mm, Leica Microsystems), Leica Historesin Embedding Kit (#702218500, Leica Microsys-

tems, Heidelburg, Germany) containing 500 mL basic resin liquid, 5 g activator powder, and 40 mL hardener.

7. Differentiation buffer for Cresyl Violet staining: Mix one volume 2.72% sodium acetate (2.72 g/100 mL) with four volumes 1.2% acetic acid (1.2 g/100 mL).

8. 10 mM sodium acetate buffer pH 6.5: 13.3 g sodium acetate trihydrate (MW=136.1) in 900 mL H_2O and pH to 6.5 with glacial acetic acid. Complete to 1 L.

9. Mounting media: Merkoglass® (Merk, Darmstadt, Germany) for paraffin-prepared sections and Mowiol (Calbiochem, San Diego, CA) for cryosections.

3. Methods

3.1. Design and Construction of Viral Vectors Encoding Small Hairpin RNA

Bipartite cassettes containing green fluorescent protein (GFP) and an shRNA sequence under control of the H1 promoter are cloned into the pSIN-W-PGK and pAAV-CMV-MCS genomic plasmids using standard molecular biology techniques to generate pSIN-PGK:GFP-H1:shRNA and pAAV-CMV:GFP-H1:shRNA, respectively (**Fig. 7.1**). The shRNA sequence consists of the 19–23 bp stem sequence targeting SOD1, the loop sequence (*TTCAAGAGA*), the reverse complementary stem sequence, and the pol III transcription termination sequence (TTTTT). Examples of complementary oligos used to clone shRNA sequence (from Raoul et al. *(7)*) are given below. Observe the 5′ overhangs that facilitate ligation into restriction-digested DNA.

5′ GATCCCC AAGGATGAAGAGAGGCATG *TTCAAGAGA* CATGCCTCTCTTC
     ATCCTT TTTTTGGAAA 3′
3′ GGG TTCCTACTTCTCTCCGTAC AAGTTCTCA GTACGGAGAGAAGTAG
     GAA AAAAACCTTTGATC 5′

The design of shRNA to target SOD1 (or any gene for that matter) is not trivial. Several algorithms have been published that can predict the silencing efficacy of siRNA sequences *(10–15)* and these have been used to create multiple online programs capable of identifying effective siRNA targets within a selected gene. Unfortunately, these algorithms for siRNA design have little or no efficacy at predicting shRNA knockdown outcome *(16)*. However, such software still offers an easy method to identify oligonucleotide sequences that may then be screened for silencing efficacy in vitro.

1. Use siRNA design software (*see* **Note 2**) to identify multiple target sequences (*n* > 10) that are located across different regions of the selected gene.

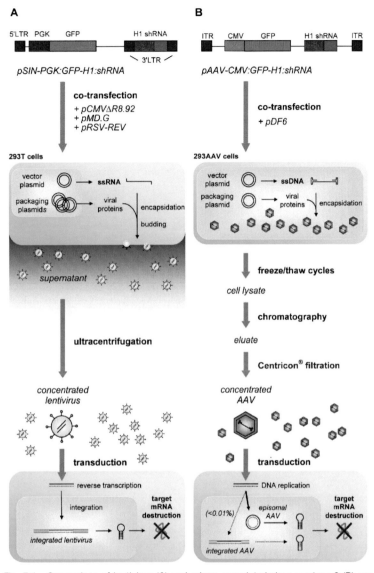

Fig. 7.1. Comparison of lentivirus (**A**) and adeno-associated virus serotype 6 (**B**) production and transduction.

2. Perform BLAST searches to eliminate those sequences that have less than three mismatches with off-target genes within the genome of the species being examined.

3. Order the oligo pairs required to generate the shRNA sequences ($n > 8$) (use template described above), anneal, and clone into an H1 expression vector.

4. Transfect the vectors into a cell line expressing the target protein and determine knockdown efficiency by Western blot, immunocytochemistry, or reverse transcription quantitative PCR (RT-QPCR).

5. Select the best two shRNA sequences for subsequent in vivo studies and design corresponding scrambled nucleotide shRNA sequences as negative controls.

3.2. Production and Titration of Recombinant Virus

Recombinant lentivirus and AAV are produced through calcium-phosphate transfection of HEK293-derived cells. HEK293 cells are human embryonic kidney cells that have been transformed by sheared adenovirus type 5 DNA (17). We use AAV-293 cells that are derived from HEK293 cells for AAV production although they produce higher viral titers (technical data sheet, #240073, Stratagene). Our lentiviral production utilizes 293T cells, which express the SV40 large T antigen allowing for episomal replication of plasmids containing an SV40 origin of replication. Indeed, the packaging and genomic plasmids used for lentivirus production contain the SV40 *ori*, leading to greater copies of plasmid and therefore increased yields of virus.

Following transient transfection, AAV particles are produced and remain within the AAV-293 cell nucleus. The cells are then lysed and the virus is purified from the lysate by high-pressure liquid chromatography on heparin columns. This is made possible by the high heparin-binding affinity of AAV6. Not all AAV serotypes, however, have an affinity for heparin and may alternatively be purified by cesium-chloride ultracentrifugation or ion-exchange chromatography (18). VSV-G-pseudotyped lentiviral vectors are packaged at the membrane following 293T transfection and delivered directly to the cell culture media. The media is collected and the virus concentrated by ultracentrifugation. **Figure 7.1** summarizes the production protocol for the two viruses.

3.2.1. Cell Culture and Transfection

1. Maintain 293 (AAV- or -T) cells with DMEM/FCS in a 37°C humidified incubator with 5% CO_2.
2. Passage cells 1:3 twice per week towards maintaining stable high-quality cell line (*see* **Note 3**).
3. Trypsinize cells from confluent tissue culture flasks the afternoon prior to transfection and centrifuge at 800 g for 5 min at 4°C to remove trypsin. Resuspend in DMEM/FCS and plate 40×10^6 cells per 500 cm^2 dish. The cells should be 70–80% confluent on the day of transfection.
4. Prepare the vector plasmid and packaging plasmid/s in 2.5 mL 0.5 M $CaCl_2$ per 500 cm^2 plate (lentivirus, 130 μg SIN-PGK:GFP-H1:shRNA, 130 μg pCMVΔR8.92, 37.5 μg pMD.G, and 30 μg pRSV-Rev; AAV, 100 μg pAAV-CMV:GFP-H1:shRNA, and 200 μg pDF6). Bring the volume to 5 mL with H_2O. Add 5 mL of 2x HBSS dropwise (one drop every other second) to the DNA/$CaCl_2$ solution while gently vortexing.

5. Incubate the solution for 10 min (*see* **Note 4**). The solution should turn opaque but no solids should be visible.
6. Add the precipitate solution gently to the cells by distributing the drops uniformly across the plate. Mix the solution in a cross 'T' manner. Return the cells to the incubator.

3.2.2. Lentivirus Isolation and Concentration

1. Remove the 293T cells from the incubator 8–15 h following transfection and aspirate the media. Replace with 70 mL of serum-free Episerf and return the dish to the incubator.
2. Collect the media (containing the virus) from the dish 48–72 h post-transfection and filter through a 0.45-μm bottle-top membrane.
3. Ultracentrifuge the filtrate at 50,000 g at 4°C for 90 min. Aspirate the supernatant and resuspend the pellet in 50 μL to 2 mL of PBS/1% BSA (depending on desired concentration). Leave the suspensions on ice for 1 h.
4. Aliquot the lentivirus and store at –80°C.

3.2.3. AAV Isolation

1. Remove the 293T cells from the incubator 6–8 h following transfection and replace media with 100 mL of serum-free Episerf. Return to incubator.
2. Collect the media from the plates 48 h following transfection and store. Trypsinize the cells and inactivate with Episerf.
3. Combine the harvested cells and collected media and centrifuge for 5 min at 800 g. Discard the supernatant and resuspend the pellet in 3 mL per 500 cm^2 plate.
4. Freeze–thaw the cells three times by cycling between a dry ice/ethanol and a 37°C water-bath.
5. Add 50 U/mL Benzonase and incubate for 37°C for 30 min.
6. Centrifuge the sample at 1,500 g for 15 min. Collect the supernatant and add 0.5% deoxycholic acid.
7. Incubate for 30 min and then centrifuge at 3,000 g for 15 min. Collect the supernatant and pass through a 0.8-μm filter.

3.2.4. AAV Purification via HPLC

1. Equilibrate the heparin column by washing with PBS for 10 min (*see* **Note 5**).
2. Load viral lysate onto the column and monitor the 280 nm absorbance. A large peak should arrive representing the non-heparin binding flow-through from the cell lysate. At this stage the AAV6 should remain bound to the column.
3. Continue to wash the column with PBS for 10 min until the peak returns to the starting absorbance or levels out in a parabolic shape.
4. Elute the virus with 0.4 M NaCl in PBS. Collect the eluate in 5 mL aliquots using the automated function of the HPLC

system. A peak smaller than the first should appear. Allow to elute until the peak levels out.

5. Strip the column of any remaining virus or sundries by washing with 1 M NaCl in PBS for 10 min.

6. Rinse the column with H_2O for 10 min and sterilize over 10 min with 20% EtOH. Leave the system including the column in 20% EtOH.

3.2.5. AAV
Concentration

1. Prepare the Centricon® Plus-20 filter column by washing with 19 mL H_2O and then with 19 mL PBS. Centrifuge at 4,000 g for 5 min.

2. Load the HPLC eluate onto the column and centrifuge at 4,000 g until approximately 1 mL of volume remains (10–15 min). Do not let the column dry out.

3. Wash the sample by adjusting the volume to 19 mL with PBS and re-centrifuging until 100–200 µL of the volume remains.

4. Aliquot the AAV and store at –80°C.

3.2.6. Virus Titration
Using FACS Analysis

Several methods are available to titrate lentivirus and AAV. Lentivirus particle content is typically performed by p24 (capsid antigen) ELISA assay (*see* **Note 6**). Similarly, AAV genomic particle content may be determined by quantitative PCR against the viral genome. These methods do not, however, quantify vector particles capable of infecting cells. When a fluorescent reporter is expressed by the vector, flow cytometry is a simple procedure that can be used to accurately titrate the vector in terms of transducing units (tu).

1. Plate 293T cells at a density of 3×10^5 cells per six-well tissue culture dish.

2. Infect the following morning with serial dilutions of the vectors. The dilutions will ensure that there are wells with a low multiplicity of infection (i.e., <15% transduction rate) that are desirable for accurate titration.

3. *Lentivirus only*: Harvest an uninfected well and determine total number of cells at time of infection (C).

4. Harvest the cells after 48 h with trypsin and inactivate with DMEM/FCS. Centrifuge for 5 min at 800 g and discard the supernatant.

5. Resuspend the pellet in 1 mL of PBS.

6. *AAV only*: Take an aliquot and quantify the total number of cells at time of analysis (D).

7. Centrifuge again for 5 min at 800 g and discard the supernatant. Resuspend the pellet in 1 mL of 4% PFA and incubate for 20 min (*see* **Note 7**).

8. Centrifuge for 5 min at 1,000 g and discard the supernatant. Resuspend the pellet in 1 mL of PBS and pass

through the cell-strainer cap of a 5 mL polystyrene round-bottom tube.

9. Perform flow cytometry to quantify the number of transduced cells.

10. As a population of transduced cells will include cells that have been transduced more than once, the percentage of transduced cells (A) will follow a Poisson's distribution in relation to transduction events per cell (B) (*formula 1*). The titer of lentivirus (T_L) in tu per microliter is calculated from transduction events in relation to the number of cells at time of infection (C) (*formula 2*, where V is the volume of virus added (μL)). The titer of AAV (T_{AAV}) in tu per microliter is calculated from transduction events in relation to the number of cells at time of analysis (D) (*formula 3*) (*see* **Note 8**).

$$\text{Formula 1}: B = \ln\left(\frac{1}{1-A}\right)$$

$$\text{Formula 2}: T_L = \frac{BD}{V}$$

$$\text{Formula 3}: T_{AAV} = \frac{BC}{V}$$

3.3. Viral-Mediated RNAi Against SOD1 in the G93A SOD1 Mouse Model of fALS

The hindlimb muscles of G93A SOD1 mice are the first muscles to display motor deficits (day 90) and become paralyzed (day 130) and invariably determine the age of death in this experimental paradigm. The hindlimb muscles and the lumbar spinal cord are therefore ideal targets for viral-mediated therapeutic silencing of SOD1. Indeed, the intraspinal and intramuscular injections described in this section specifically target the motor neurons at the level controlling hindlimb function (lumbar spinal cord). These protocols could, however, be modified to target motor neurons at different spinal cord levels to increase the therapeutic effect.

3.3.1. Intraspinal Delivery of VSV-G-Pseudotyped Lentivirus into Adult Mice

1. Place the 34G needle into the 26G needle and seal the overlap with adhesive glue. Connect the needle to the 10 μL 26G Hamilton syringe with polyethylene tubing (0.38×1.10). Replace the dead volume in the system with sterile H_2O.

2. Place the mouse in the induction chamber and anesthetize with 4% isoflurane. Remove the mouse and place on the surgery platform with the nose of the rodent in the anesthesia mask. From this point the anesthesia is maintained with 1.5% isoflurane (*see* **Note 9**). Make sure to apply and

reapply Viscotears® to the eyes to prevent ocular damage to the mouse.

3. Shave the back of the animal with an electric razor. Be careful to remove excess hairs that may contaminate the surgery. Swab the skin with Betadine® solution and allow to dry.

4. Make a 30 mm incision with the scalpel down the middle of the back. Use the tissue forceps to separate the skin as much as possible from the underlying muscle.

5. Identify the third and fourth lumbar vertebrae (L3 and L4) and mark the region by making a small scratch in the muscle surface with a scalpel (*see* **Note 10**). Make approximately 4 mm deep and 10 mm long incisions on either side of the vertebral column. Be careful not to axotomize the animal by penetrating the tissue too deep.

6. Transfer the mice to the stereotaxic frame and secure the spinal cord by placing the transverse clamps between the cranial articular process and costal process at L3 and L4 (**Fig. 7.2C,D**). The spinal column should be perpendicular the injection plane and firmly fixed such that it is unable to move upon applying downward pressure with the tissue forceps.

7. Place bandages under the mouse to support the weight of its body and reduce force exerted by the clamps.

8. Expose the vertebral column by removing the overlying muscle. This is achieved by making tiny lateral incisions down the column and carefully scraping the muscle to the side. Do not tear away or transversally cut the muscle as it may increase local inflammation.

9. Use the bone rongeurs to delicately remove the spinous process, caudal articular process, and cranial articular process at L3 and L4 (**Fig. 7.2E**) (*see* **Note 11**). The spinal cord is visible at this point.

10. Lower the needles at the desired lumbar level such that they touch the surface of the spinal cord (*see* **Note 12**). There is a small blood vessel running down the middle of the spinal cord that can help align the bilateral injection.

11. Raise the needles 1 mm from the surface and puncture holes in the dura mater with a 27G beveled needle. Place the needles back on the spinal cord surface and then slowly lower them 0.75 mm into the tissue.

12. Inject the desired amount of virus in a volume between 1 and 2 μL at 0.2 μL/min. The amount of lentivirus normally required to achieve efficient transduction is approximately 1×10^6 tu per site (≈ 100 ng p24).

13. Wait for 5 min after the injection has finished and then slowly remove the needles.

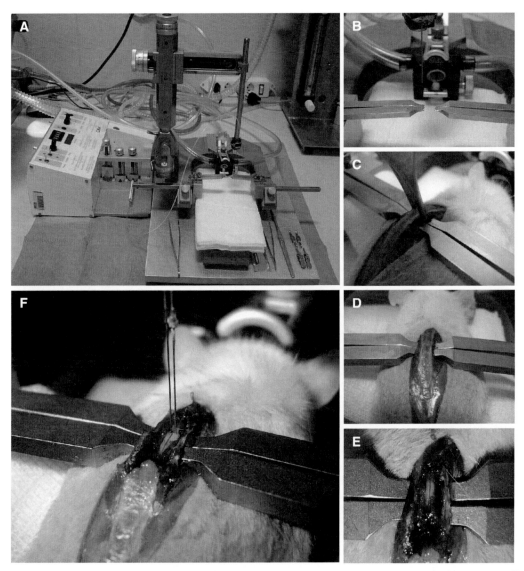

Fig. 7.2. Intraspinal injection of lentivirus into G93A SOD1 mice. (**A**) Experimental setup and required instruments. (**B**) Macrophotograph of 34G blunt needle intersecting with injection plane. (**C, D**) The vertebral column being raised with forceps to allow fixation with the transerve clamps. (**E**) The spinal cord being revealed following removal of the dorsal vertebral bones. (**F**) Macrophotograph of spinal cord being bilaterally injected with lentivirus.

14. Close the skin with 6-0 Vicryl® suture and return the animal to the housing cage (*see* **Note 13**).

3.3.2. Intramuscular Delivery of AAV6 into Adult Mice

1. Anesthetize the mouse in the induction chamber and transfer to the surgery table as described above.
2. Use an electric razor to remove the hair from the back of the lower leg and apply Betadine® solution to the surface of the skin.

3. Make an 8 mm incision with a scalpel to expose the gastrocnemius muscle. Use the tissue forceps to separate the skin from the underlying muscle.

4. Secure the legs of the animal by taping the feet to the surgery table (**Fig. 7.3B**).

5. Insert the 26G beveled needle of the 25 μL Hamilton syringe at a 45° angle into the belly of the gastrocnemius (**Fig. 7.3C,D**). Administer the desired amount of virus (diluted in 25 μL of PBS) via bolus injection (*see* **Note 14**). The amount of AAV6 required to achieve efficient transduction of motor neurons via retrograde transport is approximately 1×10^6 tu.

6. Close the skin with 6-0 Vicryl® suture and return the animal to the housing cage.

3.3.3. Intramuscular Delivery of AAV6 into Neonate Mice

1. Break off the needle from a disposable 30G insulin syringe with a needle holder by bending (not twisting) it back and forth. Connect the needle to a 26G beveled 10 μL Hamilton syringe via 20 cm of polyethylene tubing (0.28×0.61) (*see* **Note 15**).

2. Place on a clean pair of gloves and rub them in the earth of the cage of the animals. Transfer the mother of the neonates into an empty cage and remove the pup. Return the mother to the cage immediately (*see* **Note 16**).

3. Place the newborn on a latex-covered bed of crushed ice to induce hypothermic anesthesia.

4. Remove the pup from the ice and inject the desired amount of virus in 2 μL via bolus injection with the help of a second operator (**Fig. 7.3F,G**). The amount of AAV6 required to achieve efficient transduction of neonate motor neurons via retrograde transport is between 10^5 and 10^6 tu.

5. Return the pup to the cage and quickly roll it in the soiled earth. Do not return the pup directly in the nest. Instead, place the pup at least 10 cm away such that the mother is forced to return it. This will ensure that the mother does not reject the newborn.

3.4. Analysis of Transduction, SOD1 Silencing, and Neuroprotection Within the Spinal Cord

In order to adequately interpret any behavioral improvement or extended survival observed within the mouse model following viral-mediated RNAi, it is necessary to determine the transduction profile of the vector, confirm SOD1 silencing within the transduced cells, and determine whether the effects were as a result of motor neuron protection. This may be achieved via spinal cord and ventral root histology.

3.4.1. Spinal Cord Dissection

1. Administer an overdose of sodium pentobarbital and transcardially perfuse the mouse with ice-cold PBS containing 5,000 U heparin for 1 min followed by 4% PFA for 5 min.

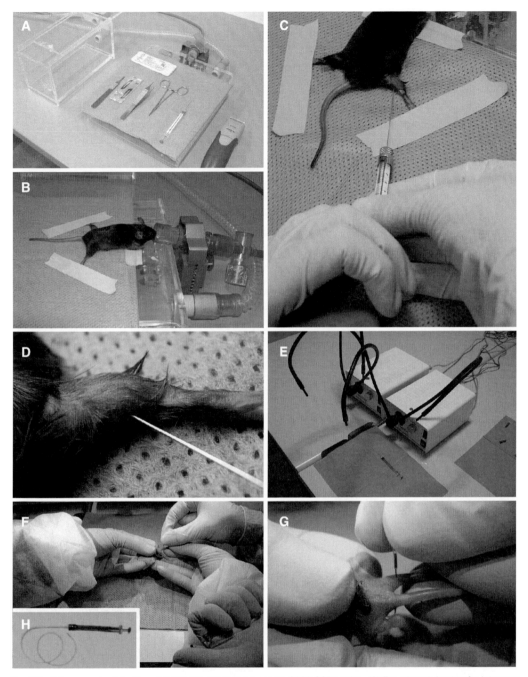

Fig. 7.3. Intramuscular injection of AAV6 into adult and neonate G93A SOD1 mice. (**A**) Experimental setup for intramuscular injection into adult mice. (**B**) Anesthetized mouse in prone position. (**C**) AAV6 being delivered into gastrocnemius muscle with 26G Hamilton syringe. (**D**) Macrophotograph of 45° needle entry into gastrocnemius. (**E**) Experimental setup for neonate intramuscular injections. (**F**) Two person coordination for ease of animal manipulation and vector injection. (**G**) Macrophotograph of needle entry into neonate gastrocnemius muscle. (**H**) 30G insulin needle connected to a 10 µL Hamilton syringe through polyethylene tubing.

2. Remove the muscle overlying the vertebral column with tissue forceps and then use the bone rongeurs to remove the vertebrae and expose the spinal cord (*see* **Note 17**).

3. Cut the spinal cord above and below the region to be dissected and then carefully remove the spinal cord with a blunt dissection probe. Use small spring scissors to cut any nerves as required to free the spinal cord.

3.4.2. Evaluation of Transduction and SOD1 Silencing; Immunohistology on Frozen Sections

1. Place the freshly dissected spinal cord in 4% PFA in PBS and postfix for 6–18 h at 4°C.

2. Transfer the spinal cords to 25% sucrose in PBS and cryoprotect overnight at 4°C. The spinal cords are ready once they have sunk to the bottom of the tube.

3. Place the spinal cord in a Peel-A-Way® disposable embedding mold containing Cryomatrix™ embedding fluid. Freeze the spinal cord by placing the mold in a beaker filled with liquid nitrogen (*see* **Note 18**).

4. Place the embedded spinal cord in a cryostat and allow to equilibrate to –20°C. Cut 20 μm transverse sections and dry for 30 min on glass slides. The sections may be stored at –80°C at this point for several months.

5. Wash slides three times with PBS over 30 min to remove cryoembedding solution and then block for 3–6 h. Incubate the sections with the primary antibodies in blocking solution overnight at 4°C. Wash the slides three times with PBS for over 45 min and then incubate with the appropriate fluorescent secondary antibodies in PBS/1% BSA for 2 h. Wash again three times with PBS for over 45 min and then rinse briefly in H_2O prior to mounting in Mowiol.

6. Examine GFP expression (corresponding to virus transduction and shRNA delivery) using either native GFP expression or enhance the signal using antibodies against GFP (1:800 goat polyclonal anti-GFP Ab with 5% NDS, 0.1% BSA, 0.1% Triton X-100 blocking solution). Analyze the GFP expression with a confocal microscope for increased resolution (**Fig. 7.4A**).

7. Quantify motor neuron transduction by counting the total number of GFP-positive cells that colocalize with VAChT-positive cells and express these as a percentage of total VAChT-positive cells (**Fig. 7.4B**) (1:800 goat polyclonal anti-GFP Ab and 1:2500 rabbit polyclonal anti-VAChT with 5% NDS, 0.1% BSA, 0.1% Triton X-100 blocking solution).

8. Examine SOD1 silencing in motor neurons/glial cells by double labeling sections from silencer and mismatch injected animals for SOD1 and GFP. SOD1 staining is achieved in TBS using either 1:1500 anti-SOD1 clone SD-G6, 1:1500 anti-SOD1 clone G215-1, or 1:100 polyclonal goat anti-SOD1 Ab in combination with 1:800 rabbit polyclonal

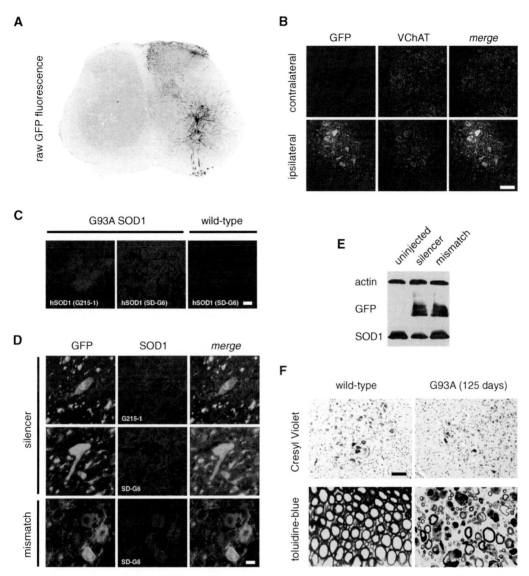

Fig. 7.4. Analysis of transduction, SOD1 silencing, and neuroprotection following viral-mediated-RNAi in the G93A SOD1 mouse. (**A**) GFP fluorescence of lumbar spinal cord following unilateral intramuscular injection of AAV6. (**B**) Double labeling demonstrating colocalization between GFP and the motor neuron marker, VAChT, in the ventral horn. Scale bar, 80 μm. (**C**) Specificity of SOD1 staining with two different antibodies comparing G93A SOD1 and wild-type mice. Scale bar, 20 μm. (**D**) Immunostaining for SOD1 and GFP in the ventral horn of mice injected with lentivirus expressing either the silencer or mismatch shRNA against SOD1. Scale bar, 20 μm. (**E**) Immunoblot evaluation of vector-mediated silencing in the lumbar spinal cord following direct intraspinal delivery. (**F**) Cresyl Violet and toluidine-blue sections of ventral horn spinal cord and ventral roots, respectively, from wild-type and G93A SOD1 mice at 125 days of age. Scale bar, 100 μm. (**C–F** reproduced fromk Ref. *(7)* with permission from Nature Publishing Group).

anti-GFP Ab with 4% NDS, 4% BSA, 0.1% Triton X-100 blocking solution. Merge images from the GFP and SOD1 channels to best illustrate SOD1 knockdown (**Fig. 7.4C**) (*see* **Note 19**).

3.4.3. Evaluation of Neuroprotection; Motor Neuron Counts on Paraffin-Embedded Sections

1. Place the freshly dissected spinal cord in 4% PFA in PBS and postfix for 6–18 h at 4°C.
2. Prepare spinal cord for paraffin inclusion by rinsing two times with PBS for 15 min; two times with 25% EtOH/75% PBS for 10 min; two times with 50% EtOH/50% PBS for 10 min; two times with 70% EtOH for 20 min.
3. Perform the following cycles using an automatic embedding machine: 70% EtOH for 90 min; 80% EtOH for 90 min; 96% EtOH for 90 min; 100% EtOH for 90 min; toluol for 90 min; paraffin for 2 h at 60°C.
4. Embed the spinal cord in the embedding mold and solidify the paraffin at 4°C.
5. Cut 8 μm transverse spinal cord sections and place on a 1% gelatin solution heated at 42°C. Mount the sections onto glass slides and dry for 30 min.
6. Deparaffinize the slides for heating for 1 h at 60°C and place immediately in toluol. Leave for 10 min and transfer to a fresh batch of toluol for a further 5 min.
7. Rehydrate the slides by rinsing: two times with 100% EtOH for 2 min; one time with 96% EtOH for 2 min; one time with 80% EtOH for 2 min; one time with 70% EtOH for 2 min; one time H_2O for 5 min.
8. Place slides for 1 min in 0.5% Cresyl Violet in differentiation buffer and then leave 30 s in differentiation buffer alone. Wash two times for 30 s in 100% EtOH and then two times for 30 s in toluol.
9. Mount the slides in Merkoglass® and allow to dry overnight.
10. Count only large-sized cells (>25 μm) (motor neurons) with a visible nucleolus on every fifth section across the desired region of spinal cord. Examples of stained sections are shown in **Fig. 7.4F** (top panel).

3.4.4. Evaluation of Neuroprotection; Myelinated Fiber Counts on Resin-Embedded Sections

One pathological characteristic of ALS observed in the clinic and in rodent models is the demyelination of large axons (motor neurons) in the spinal cord ventral roots. Quantifying the numbers of myelinated fibers therefore provides another measure of protection from motor neuron degeneration.

1. Dissect the ventral roots from the dissected spinal cord under a stereomicroscope and postfix in 4% PFA/2.5% glutaraldehyde for 6–18 h at 4°C.

2. Incubate ventral roots in the dark for 4 h in 2% osmium tetroxide, 50 mM cacodylate (*see* **Note 1**).

3. Wash the ventral roots rapidly in H_2O and dehydrate with 80% EtOH for 1 h, 96% EtOH for 1 h, 96% EtOH for 1 h, 96% EtOH/infiltration solution (1:1) for 1 h, and then infiltration solution overnight at 4°C.

4. Embed ventral roots in 2 × 15 mm molds by mixing the infiltration solution with the hardener and allow to set.

5. Cut 2 μm transverse sections using an ultramicrotome and place sections on a glass slide (*see* **Note 20**).

6. Place the slides on a heating plate at 120°C for 10 min and let dry.

7. Incubate the slides in 0.5% Toluidine Blue O, 10 mM sodium acetate pH 6.5 for 1 h.

8. Wash slides two times for 30 s in H_2O, two times for 60 s in 100% EtOH, two times for 60 s in toluol, and mount in Merkoglass®.

9. Count the numbers of small (<5 μm) and large (≥5 μm) myelinated motor fibers in the ventral roots. The lipid-rich myelin is stained black by the lipophilic osmium tetroxide. Examples of stained sections are shown in **Fig. 7.4F** (bottom panel).

4. Notes

1. Paraformaldehyde and osmium tetroxide are extremely hazardous. Observe all warning and precautions listed on the MSDS for these compounds.

2. The Whitehead Institute for Biomedical Research offers an excellent siRNA target designer that is free (http://jura.wi.mit.edu/bioc/siRNAext/). Registration by email is required.

3. Viral production yield may decrease as passage number increases in the 293-derived cells. We recommend producing a cell bank with cells at low passage to allow regular thawing of new vials. Cell may be frozen in 80% serum and 10% dimethyl sulfoxide (DMSO) and stored in liquid nitrogen.

4. Unless otherwise stated, all incubations, washes, drying steps, and centrifugations are performed at room temperature.

5. New heparin columns should be washed with PBS for 10 min, followed by 0.4 M NaCl in PBS, followed again by PBS for 10 min.

6. There are many commercially available kits to titrate p24 antigen. We have previously used with success the

RETROtek HIV-1 p24 Antigen ELISA kit (#0801200, ZeptoMetrix Corporation, Buffalo, NY).

7. Gently flick the tube one or two times prior to adding the PFA to disperse the pellet. This reduces the number of cell clumps observed following addition of the fixative.

8. The reason for the difference between the calculations results from the fact that lentivirus integrates whereas AAV does not. AAV genomes remain mostly episomal within the nucleus of cells and will be diluted as the cells divide.

9. The percentage of isoflurane required to induce and maintain anesthesia may vary between strain, age, and sex of the rodents as well as the apparatus being used.

10. L3 and L4 are identified by using the index finger and thumb to localize the pelvic bone at the hip level which corresponds to L6. The vertebrae are then counted backwards to localize the desired level.

11. Knowledge of the anatomy of mouse vertebrae will help at this point. We recommend consulting a mouse anatomy book prior to performing this procedure.

12. The use of a stereomicroscope or magnifying glass will aid visualization of the injection.

13. 1 g/L of effervescent paracetamol can be added to the water supply of the animal in order to minimize postoperative pain and hasten recovery.

14. The muscle should expand in volume following virus delivery. This is easily observed visually and indicates a successful injection.

15. We have found that this setup is the most practical for neonate injections. One hand can be used to hold and insert the needle into the muscles while the other hand remains free to inject the virus with the syringe. This increases the dexterity of the manipulation.

16. There is always a chance that experimental neonate manipulations may provoke the mother to kill the newborns. The animal handling method described in this section is the best in our experience to avoid this.

17. Removing the vertebrae is the most difficult procedure in spinal cord dissection. Great care must be taken to avoid damaging the spinal cord that, despite being fixed, remains extremely delicate. After clearing the muscle from either side of the vertebra, use the bone rongeurs to carefully 'crack' the column at each vertebral level (in the same way as one would crack open a walnut). *Be careful not to apply so much force that the vertebrae collapse onto the underlying spinal cord.* After this step, use the bone rongeurs to carefully remove the piece of bone comprising the spinous process and cranial articular process at the level below the region of spinal cord that is to be dissected. This action

will be transversal to the column. Next, remove the piece of bone comprising the cranial articular process with the caudal articular process. This action will be lateral to the column. Repeat these actions (transversal, lateral, transversal, etc.) moving up the spinal column until sufficient spinal cord is revealed.

18. Follow the appropriate safety procedures for handling liquid nitrogen as specified by your institution.

19. SOD1 silencing can also be demonstrated via Western blot (**Fig. 7.4D**) or RT-QPCR (not shown).

20. Placing a drop of water on the glass slide prior to placement of the resin section will help to unwrinkle and flatten the section.

Acknowledgments

The authors would like to thank Philippe Colin, Vivianne Padrun, Fabienne Pidoux, and Christel Sadeghi for their expert technical assistance. This work was supported by the ALS Association (ALSA), the Sixth Research Framework Programs of the European Union, Project RIGHT (LSHB-CT-2004 005276), Project NeuroNE (LSHM-CT-2004-512039), and the Bruno and Ilse Frick Foundation.

References

1. Bruijn, L.I., Miller, T.M. and Cleveland, D.W. (2004) Unravelling the mechanisms involved in motor neuron degeneration in ALS. *Annu. Rev. Neurosci.* **27**, 723–749.

2. Dykxhoorn, D.M., Novina, C.D. and Sharp, P.A. (2003) Killing the messenger: short RNAs that silence gene expression. *Nat. Rev. Mol. Cell Biol.* **4**, 457–467.

3. Saito, Y., Yokota, T., Mitani, T., Ito, K., Anzai, M., Miyagishi, M., Taira, K. and Mizusawa, H. (2005) Transgenic small interfering RNA halts amyotrophics lateral sclerosis in a mouse model. *J. Biol. Chem.* **280**, 42826–42830.

4. Smith, R.A., Miller, T.M., Yamanaka, K., Monia, B.P., Condon, T.P., Hung, G., Lobsiger, C.S., Ward, C.M., McAlonis-Downes, M., Wei, H., Wancewicz, E.V., Bennett, C.F. and Cleveland, D.W. (2006) Antisense oligonucleotide therapy for neurodegenerative disease. *J. Clin. Invest.* **116**, 2290–2296.

5. Zamore, P.D., Tuschl, T., Sharp, P.A. and Bartel, D.P. (2000) RNAi: double stranded RNA directs the ATP-dependent cleavage of mRNA at 21 to 23 nucleotide intervals. *Cell* **101**, 25–33.

6. Ralph, G.S., Radcliffe, P.A., Day, D.M., Carthy, J.M., Leroux, M.A., Lee, D.C., Wong, L.F., Bilsland, L.G., Greensmith, L., Kingsman, S.M., Mitrophanous, K.A., Mazarakis, N.D. and Azzouz, M. (2005) Silencing mutant SOD1 using RNAi protects against neurodegeneration and extends survival in an ALS model. *Nat. Med,* **11**, 429–433.

7. Raoul, C., Abbas-Terki, T., Bensadoun, J.C., Guillot, S., Haase, G., Szulc, J., Henderson, C.E. and Aebischer, P. (2005) Lentiviral-mediated silencing of SOD1 through RNA interference retards disease onset and progression in a mouse model of ALS. *Nat. Med.* **11**, 423–428.

8. Boillee, S., Vande Velde, C. and Cleveland, D.W. (2006) ALS: a disease of motor

neurons and their nonneuronal neighbors. *Neuron* **52**, 39–59.

9. Miller, T.M., Kaspar, B.K., Kops, G.J., Yamanaka, K., Christian, L.J., Gage, F.H. and Cleveland, D.W. (2005) Virus-delivered small RNA silencing sustains strength in amyotrophic lateral sclerosis. *Ann. Neurol.* **57**, 773–776.

10. Reynolds, A., Leake, D., Boese, Q., Scaringe, S., Marshall, W.S. and Khvorova, A. (2004) Rational siRNA design for RNA interference. *Nat. Biotechnol.* **22**, 326–330.

11. Saetrom, P. and Snove, O.J. (2004) A comparison of siRNA efficacy predictors. *Biochem. Biophys. Res. Commun.* **321**, 247–253.

12. Takasaki, S., Kotani, S. and Konagaya, A. (2004) An effective method for selecting siRNA target sequences in mammalian cells. *Cell Cycle* **3**, 790–795.

13. Ui-Tei, K., Naito, Y., Takahashi, F., Haraguchi, T., Ohki-Hamazaki, H., Juni, A., Ueda, R. and Saigo, K. (2004) Guidelines for the selection of highly effective siRNA sequences for mammalian and chick RNA interference. *Nucleic Acids Res.* **32**, 936–948.

14. Hsieh, A.C., Bo, R., Manola, J., Vazquez, F., Bare, O., Khvorova, A., Scaringe, S. and Sellers W.R. (2004) A library of siRNA duplexes targeting the phosphoinositide 3-kinase pathway: determinants of gene silencing for use in cell-based screens. *Nucleic Acids Res.* **32**, 893–901.

15. Amarzguioui, M. and Prydz, H. (2004) An algorithm for selection of functional siRNA sequences. *Biochem. Biophys. Res. Commun.* **316**, 1050–1058.

16. Taxman, D.J., Livingstone, L.R., Zhang, J., Conti, B.J., Iocca, H.A., Williams, K.L., Lich, J.D., Ting, J.P. and Reed, W. (2006) Criteria for effective design, construction, and gene knockdown by shRNA vectors. *BMC Biotechnol.* **6**, 7.

17. Graham, F.L., Smiley, J., Russell, W.C. and Nairn, R. (1977) Characteristics of a human cell line transformed by DNA from human adenovirus type 5. *J. Gen. Virol.* **36**, 59–74.

18. Grimm, D., Kay, M.A. and Kleinschmidt, J.A. (2003) Helper virus-free, optically controllable, and two-plasmid-based production of adeno-associated virus vectors of serotypes 1 to 6. *Mol. Ther.***7**, 839–850.

Chapter 8

Studying Autoimmunity by In Vivo RNA Interference

Stephan Kissler

Abstract

The occurrence of autoimmunity is strongly associated with multiple gene variants that predispose individuals to disease. The identification of the gene polymorphisms that modulate disease susceptibility is key to our understanding of disease etiology and pathogenesis. While genetic studies in humans have uncovered several associations and have provided possible candidate genes for further study, the use of animal models is indispensable for detailed functional studies. In order to facilitate the genetic manipulation of experimental models of autoimmunity, we employ lentiviral transgenesis in combination with RNA interference (RNAi). This approach bypasses the need for targeted mutagenesis of embryonic stem cells and/or backcrossing of genetically modified animals onto the relevant genetic background. Lentiviral RNAi offers several advantages compared to conventional transgenesis or knockout technology, and these, as well as the technique's weaknesses, are discussed herein.

Key words: Lentivirus, RNAi, autoimmunity, transgenesis, mouse, embryo.

1. Introduction

RNAi has found widespread use in many laboratories, and has rapidly become a method of choice for genetic manipulation of transformed cell lines as well as whole organisms such as *Caenorhabditis elegans* and even *Drosophila melanogaster*. Its use in mammalian organisms is still limited by one main hurdle: specific and long-lasting delivery of oligonucleotides to cells within a living animal. Just seven years ago, Baltimore and colleagues demonstrated that lentivirus could be employed to deliver a transgene to single-cell mouse embryos that subsequently developed into transgenic adult animals *(1)*. That same year, Brummelkamp et al. demonstrated that small interfering RNAs (siRNAs) could

John F. Reidhaar-Olson and Cristina M. Rondinone (eds.), *Therapeutic Applications of RNAi: Methods and Protocols, vol. 555*
© Humana Press, a part of Springer Science+Business Media, LLC 2009
DOI 10.1007/978-1-60327-295-7_8 Springerprotocols.com

be expressed from a short-hairpin motif (shRNA) incorporated within a retroviral vector *(2)*. The following year, several groups had combined these two technologies to generate transgenic mice in which a particular gene was silenced by RNAi *(3, 4)*. Since then, lentiviral vectors have been refined and modified to accommodate the inducible expression of shRNA sequences within transgenic animals *(5, 6)*.

The study of autoimmunity, and particularly of type 1 diabetes (T1D), requires the use of very specific mouse strains. In the case of T1D, the best and most widely studied model is the nonobese diabetic (NOD) mouse. This mouse model closely resembles human T1D, and shares not only pathophysiological aspects, but also genetic susceptibility loci with human disease. The availability of the NOD mouse has allowed several disease-associated human genes to be confirmed and studied in more detail in the NOD model. However, the NOD mouse is notoriously refractory to genetic manipulation. Despite prolonged efforts, no NOD-derived embryonic stem cell line is yet available that would allow direct manipulation of the NOD genome by targeted mutation. Even conventional transgenesis by pronuclear injection is more challenging in this strain, such that the generation of mutant NOD mice often requires lengthy backcrossing of mutant animals from a different strain onto the NOD background. We have now used lentiviral transgenesis to silence genes directly in the NOD embryo *(7)*, thereby greatly reducing the time required to generate new experimental models and to test the involvement of candidate genes in the disease process.

2. Materials

2.1. Cloning of Oligonucleotides into the Lentiviral Vector

1. Synthetic oligonucleotides (ordered commercially).
2. Lentiviral vector (e.g. pLB, available from Addgene).
3. Restriction enzyme (for pLB: *Hpa*I, *Xho*I, *Xba*I).
4. T4 Polynucleotide kinase with T4 PNK buffer.
5. T4 ligase with ligase buffer.
6. Competent bacteria (e.g. Novablue (Novagen)).
7. LB medium, supplemented with 50 µg/ml ampicillin.
8. LB-agar plates, supplemented with 50 µg/ml ampicillin.

2.2. Generation of Lentivirus

1. DMEM supplemented with penicillin (100 Units/ml) and streptomycin (100 µg/ml), with and without 5 or 10% fetal bovine serum (FBS).
2. HEK 293 cells.
3. Transfection reagent (e.g. FuGene6 (Roche) or Polyfect (Qiagen)).

4. Lentiviral packaging plasmids (pMDLg/pRRE and pRSV-Rev, available from Addgene).

5. Envelope plasmid (pCMV-VSVg, available from Addgene).

6. Ultracentrifuge tubes (40 ml capacity).

7. Sterile PBS.

2.3. Embryo Generation, Manipulation and Injection

1. Embryo donor mice (females from strain of choice).
2. Male stud mice (from strain of choice).
3. Pregnant Mare Serum (PMS) gonadotropin (Sigma-Aldrich) and human Chorionic Gonadotropin (hCG) (Sigma-Aldrich).
4. M16 medium, M2 medium, mineral oil, hyaluronidase (all from Sigma-Aldrich).
5. Microinjection set-up.

2.4. Genotyping of Transgenic Animals

1. (Optional) Multi-LED Flashlight with blue filter (e.g. Lee Filters #47B Tricolor Blue) and goggles with yellow filter (e.g. Lee Filters #12).
2. Fluorescent microscope with appropriate filters (for GFP visualization).
3. Flow cytometer.
4. PBS supplemented with 5 mM EDTA.
5. Red blood cell lysis buffer (ACK: 0.15 M NH_4Cl, 1 mM $KHCO_3$, 0.1 mM Na_2EDTA, pH 7.2–7.4).
6. PBS supplemented with 1% FBS.

3. Methods

The principle of the procedure consists in designing a lentiviral vector that contains an shRNA-encoding construct directed against the gene of interest. The lentiviral vector is packaged in VSVg-pseudotyped viral particles that are microinjected into the perivitelline space of single-cell embryos. The embryos are reimplanted within 24 h of injection, and develop into fully transgenic animals in which the gene of interest is silenced by RNAi.

3.1. Cloning of Oligonucleotides into the Lentiviral Vector

All procedures described below apply to cloning into the pLB vector. Using a different vector will require adaptation of the procedure to incorporate the relevant design and restriction sites.

1. Identify suitable 19-nt target sequences within the gene of interest, using one of the freely available algorithms.
2. The 19-nt target sequence is incorporated into an shRNA design (3) as follows:

Forward strand: T(N+)$_{19}$TTCAAGAGA(N–)$_{19}$TTTTTC

Reverse strand: TCGAGAAAAA(N+)$_{19}$TCTCTTGAA(N−)$_{19}$A

where (N+)$_{19}$ denotes the target sequence and (N−)$_{19}$ denotes the antisense complementary sequence to the target sequence.

3. Order the forward and reverse strands as unmodified, desalted oligonucleotides at a small (10 or 25 nmol) scale. Resuspend oligonucleotides at 100 pmol/μl (*see* **Note 1**).

4. Phosphorylate 5 μl of each oligonucleotide in a total volume of 10 μl with T4 polynucleotide kinase in the appropriate buffer for 1 h at 37°C.

5. Combine the two phosphorylation reactions into one PCR tube (total 20 μl) and anneal the two strands by heating to 94°C for 2 min, followed by 10 min at 70°C, and cooling down to 4°C slowly (1°C/s). Dilute the annealed dsDNA in a total of 500 μl.

6. Digest 3 μg of the lentiviral vector (pLB) using *Hpa*I and *Xho*I. Dephosphorylate the vector (optional) and purify the digested vector (using a commercially available PCR-purification kit, for example) into 30 μl.

7. Set up a ligation with 5 μl digested vector together with 1 μl annealed oligonucleotide and 1 μl T4 ligase in 20 μl using the appropriate ligase buffer. Leave at room temperature for 2–3 h, and transform chemically competent bacteria with 0.5–1 μl of the ligation reaction.

8. Colonies are screened after selection on LB-agar/ampicillin plates and growth in LB/ampicillin medium by restriction digest of the purified plasmid DNA using *Xba*I and *Xho*I. Visualize the digest on a 1.5% agarose gel: the empty vector should yield a band at approximately 330 bp, while a vector with the correct insert should have a band at 385 bp. Positive colonies should then be verified by sequencing, as mutations in the shRNA-encoding sequence do occasionally occur during the cloning process.

3.2. Generation of Lentivirus

1. Prepare HEK 293 cells the day prior to transfection: plate out approximately 8–10 million cells in a 15-cm tissue culture plate or, alternatively, in a 175 cm^2 flask in 20 ml DMEM with 10% FBS.

2. The next day, prepare the transfection mixture as follows: 20 μg lentiviral vector, 4 μg pMDL-g/p RRE, 3 μg pRSV-Rev, 3 μg pCMV-VSVg, together with 60 μl transfection reagent in serum-free DMEM. Incubate at room temperature for at least 20 min, and pipette onto the cells (*see* **Note 2**).

3. The day following transfection, replace the medium with DMEM 5% FBS (*see* **Note 3**).

4. Collect culture supernatant at 48 h and store at 4°C, replacing it with 20 ml fresh medium (5% FBS).

5. Collect culture supernatant at 72 h, and combine with the 48 h supernatant.

6. Spin the combined supernatants at 2500 rpm for 10 min to pellet cell debris. Filter the supernatant through a 0.45-μm sterile filter and ultracentrifuge at 25,000 rpm for 90 min.

7. Carefully remove the supernatant by aspiration. Add 100 μl sterile PBS, taking care not to disrupt the pellet, if one is visible (*see* **Note 4**). Cover tubes with parafilm and leave undisturbed at 4°C overnight.

8. The following day, hold the ultracentrifuge tube at an angle and pipette the PBS onto the opposite tube wall, so it flows over the pellet. Repeat 10–20 times, then transfer the virus solution into 1.5-ml tubes (usually as 10–20 μl aliquots) and freeze at −80°C (*see* **Note 5**).

9. To titer the virus, prepare HEK 293 cells (400,000/well) in a six-well plate with DMEM/10% FBS. The following day, dilute 1 μl virus in 1 ml DMEM/10% FBS, and transfer 100 μl of this mixture into 1 ml of DMEM/10% FBS to generate a 1:10 dilution. Repeat the dilution to obtain a 1:100 dilution. Replace the medium in three wells of the previously prepared HEK 293 cells with the three virus dilutions and in one well with virus-free DMEM/10% FBS. Assuming that approximately 800,000 cells were present (doubling of the initial cell number) at the time of infection, the proportion of infected cells can be measured 48 h post-infection by flow cytometry, and the virus titer extrapolated from the percentage infected (GFP-positive) cells (*see* **Note 6**).

10. The virus can be used to infect a variety of cells, and validation of the chosen shRNA sequences can be performed by measuring mRNA or protein level in an adequate cell type after infection (*see* **Note 7** for an alternative validation method).

3.3. Embryo Generation, Manipulation and Injection

1. Superovulate donor females of the chosen mouse strain by intraperitoneal injection of 5 I.U. PMS on day 2, and then with 5 I.U. hCG 47 h later, on day 0. The females are mated with stud males of the same strain immediately after hCG injection (*see* **Note 8**).

2. Collect embryos on day 1, shortly after fertilization: the oviduct is removed, placed into M2 medium supplemented with hyaluronidase (300 μg/ml), and embryos are freed by tearing of the oviduct. Hyaluronidase acts to free the embryos from surrounding cumulus cells.

3. Wash embryos in M2 medium, and transfer them into a drop of M16 medium to keep at 37°C in the incubator while preparing the injection needle.

4. Thaw a virus aliquot on ice, and spin at 2,000 rpm for 2 min to pellet any larger debris. Backload the microinjection needle with 3 μl of virus (*see* **Note 9**).

5. Transfer embryos into a drop of M2 medium overlayed with mineral oil for injection. Each embryo is injected with sufficient virus solution to achieve a visible expansion of the zona pellucida. Take care not to damage the actual zygote, as the virus is injected between the zona pellucida and the zygote, in the perivitelline space.

6. Transfer injected embryos into M16 medium, place in incubator at 37°C. Reimplant embryos either the same or following day into the oviduct of pseudopregnant recipients (*see* **Note 10**).

3.4. Genotyping of Transgenic Animals

Once potentially transgenic animals are born (after the usual 20–21 days gestation), three simple methods can be used to test animals for expression of the GFP marker gene.

1. The earliest possible genotyping can be performed using a flashlight and goggles with appropriate filters (see materials for filter details, and **Note 11** for further information on building your own set of flashlight and goggles). This method can be applied from the first day post-partum, and is particularly effective in albino or agouti mice with light skin. The animals are illuminated with the filtered light and observed through the filter-goggles within their cage.

2. Alternatively, a tissue biopsy (e.g. tail tip) can be obtained and observed under a fluorescent microscope equipped with appropriate filters (*see* **Note 12**).

3. For quantitation of cells that express the marker gene, a blood sample can be collected for analysis by flow cytometry. For this purpose, a small amount of blood (50–100 μl) is collected in a microfuge tube filled with 200 μl PBS/EDTA buffer to prevent clotting. The red blood cells are lysed with ACK buffer (2–3 ml for 3 min), and then washed twice with PBS/1% FBS. Cells are analysed by flow cytometry to quantify the proportion of GFP-positive cells. Additionally, specific cell-surface markers can be labelled with the relevant fluorescent antibodies to distinguish individual cell populations, if required (*see* **Note 13**).

3.5. Summary of Advantages of Lentiviral RNAi in Mice

1. Using lentiviral transgenesis bypasses the limitations imposed by the availability of embryonic stem cells from a small number of mouse strains, or even species. Lentiviral transgenesis combined with RNAi allows the downregulation of genes directly in many specialized mouse models.

2. RNAi in itself has the advantage of allowing silencing of individual splice variants by targeting unique exon boundaries, for example. This was previously not possible in many cases using conventional knockout technology. In addition, RNAi can be used to downregulate, rather than completely eliminate, a particular gene product to mimic physiological polymorphisms encountered in human disease-associated genes.

3. The use of RNAi also enables simpler tissue-specific downregulation of genes, using appropriate promoters for the expression of shRNA constructs *(8)*.

4. Lentiviral embryo injection is technically easier and more efficient than the targeting of embryonic stem cells and their injection into blastocysts or pronuclear injection of DNA.

3.6. Summary of Disadvantages of Lentiviral RNAi in Mice

1. Expression of lentiviral transgenes is very often variegated, necessitating careful analysis of founders.

2. Lentiviral integration in the embryonic genome is random, and can result in multiple integrants spread throughout the genome. These are segregated during breeding, sometimes leading to changes in expression between founder and offsprings.

3. Downregulation of gene expression by RNAi is seldom complete. Conventional knockout technology is therefore still a more reliable approach in most instances.

RNAi in vivo by lentiviral transgenesis has opened new possibilities and should facilitate the study of specialized disease models such as the NOD model of T1D. But this method is unlikely to entirely replace conventional pronuclear transgenesis or knockout technology, as these approaches are, in many instances, still more reliable and appropriate. Nevertheless, lentiviral RNAi complements conventional methods and should, as it becomes more established, prove very useful in the genetic manipulation of mammalian disease models.

4. Notes

1. We initially ordered HPLC-purified oligonucleotides, but found desalted oligonucleotides to be of sufficient quality.

2. We have used both FuGene6 (Roche) and Polyfect (Qiagen) with success. At this large scale of transfection, we find that a twofold excess of transfection reagent over DNA amount is usually sufficient for very high transfection efficiency and production of high virus titers.

3. Replacing the medium with a lower percentage FBS improves the quality of the virus concentrate by reducing

debris and impurities that can be a major problem during microinjection.

4. The virus pellet itself is not visible, and if a light-brown pellet is visible, it is likely due to impurities and debris. As a general gauge of quality of the virus concentrate, the less you see, the cleaner the virus.

5. We usually flash-freeze virus aliquots in liquid nitrogen prior to transfer to $-80°C$, but it is debatable whether this is a necessary procedure to safeguard the virus against degradation during freezing.

6. This titration method yields an approximate measure of the minimum number of infectious particles present in the virus stock. We always aim to use virus preparations with no less than $2-3 \times 10^8$ infectious particles/ml for embryo injection. As a general guideline, good virus stock should infect >50% of the HEK 293 cells in the well containing 1 μl virus.

7. As an alternative to measuring endogenous gene expression in a cell line, we routinely use a luciferase reporter assay to validate shRNA sequences. The relevant cDNA (e.g. from the OpenBiosystems collection) is cloned into the 3′ UTR of the *Renilla* luciferase gene in the dual-luciferase reporter vector psiCheck2 (Promega), and knockdown efficiency is measured in a dual-luciferase assay after co-transfection of the lentiviral vector and reporter plasmid into HEK 293 cells.

8. While some laboratories use exclusively female donors aged 3–5 weeks, others use animals of 6–8 weeks. The success of the superovulation procedure is age-dependent as well as strain-dependent, and adjustments to hormone quantities and donor age have to be made on a case-by-case basis for optimal results.

9. The injection needles used for virus injection are considerably larger than needles used for pronuclear injection. The virus concentrate can be viscous and contain larger debris that will clog small openings, one of the main technical problems encountered during this procedure. At the same time, a larger opening inflicts more damage to the embryo, and it is a matter of experience to determine what size needle works best.

10. The typical survival rate of injected embryos, as measured by two-cell development the day following injection, is up to 90% for B6 embryos. However, survival is strain-dependent, and we found NOD embryos, for example, to be much less robust. We therefore reimplant NOD embryos on the day of injection, while B6 embryos can be reimplanted the following day.

11. While sets of flashlights and goggles equipped with appropriate filters for GFP visualization are available commercially, we find that "home-made" devices are entirely adequate and much cheaper. We use a 16-LED flashlight with blue LEDs, supplemented with a photographic filter to further narrow the excitation light wavelength range (the polyester filter is cut to size to fit the flashlight). We similarly use a yellow filter cut to fit onto laboratory safety goggles, through which we observe transgenic animals. This allows visualization of GFP in the skin of animals at ambient light, or for even better results, in a dark room.

12. The second method is particularly useful when animals are held off-site, such that tail-biopsies can be shipped between laboratories. The GFP is stable for several days, if not weeks, in the tissue sample, so these can be shipped at room temperature. The GFP can be visualized directly through an eppendorf tube wall, making it a very fast and easy procedure.

13. Depending on the integration site of the transgene, we have found widely variable expression levels. These can also change after breeding; for example, if the transgene is located in an imprinted locus or if it is present in multiple copies throughout the genome. Variegated expression is a very common occurrence in lentiviral transgenics, which makes it a less reliable method than conventional knockout technology or pronuclear transgenesis. However, the efficiency of transgenesis can be higher than 50% under optimal conditions, such that the number of positive founders compensates to some extent for the wide variability in transgene expression.

References

1. Lois C., Hong E.J., Pease S., Brown E.J. and Baltimore D. (2002) Germline transmission and tissue-specific expression of transgenes delivered by lentiviral vectors. *Science* 295: 868–872.

2. Brummelkamp T.R., Bernards R. and Agami R. (2002) A system for stable expression of short interfering RNAs in mammalian cells. *Science* 296: 550–553.

3. Rubinson et al. (2003) A lentivirus-based system to functionally silence genes in primary mammalian cells, stem cells and transgenic mice by RNA interference. *Nat. Genet.* 33: 401–406.

4. Tiscornia G., Singer O., Ikawa M. and Verma I.M. (2003) A general method for gene knockdown in mice by using lentiviral vectors expressing small interfering RNA. *Proc. Natl. Acad. Sci. USA* 100: 1844–1848.

5. Ventura A., Meissner A., Dillon C.P., McManus M.T., Sharp P.A., Van Parijs L., Jaenisch R. and Jacks T. (2004) Cre-lox conditional RNA interference from transgenes. *Proc. Natl. Acad. Sci. USA* 101: 10380–10385.

6. Szulc J., Wiznerowicz M., Sauvain M.-O., Trono D., and Aebischer P. (2006) A versatile tool for conditional gene expression and knockdown. *Nat. Methods* 3: 109–116.

7. Kissler S., Stern P., Takahashi K., Hunter K., Peterson L.B. and Wicker L.S. (2006) In vivo

RNA interference demonstrates a role for Nramp1 in modifying susceptibility to type 1 diabetes. *Nat. Genet.*38: 479–483.

8. Stegmeier F., Hu G., Rickles R.J., Hannon G.J., and Elledge S.J. (2005) A lentiviral microRNA-based system for single-copy polymerase II-regulated RNA interference in mammalian cells. *Proc. Natl. Acad. Sci. USA*102: 13212–13217.

Chapter 9

Transcriptional Gene Silencing Using Small RNAs

Daniel H. Kim and John J. Rossi

Abstract

RNA interference is a potent gene silencing pathway initiated by short molecules of double-stranded RNA. Small interfering RNAs (siRNAs) with full sequence complementarity to mRNAs induce cleavage of their target transcripts in the cytoplasm. Recent evidence has shown, however, that siRNAs can also function in the nucleus of mammalian cells to affect changes in chromatin structure. When targeted to promoter regions, siRNAs load into the effector protein Argonaute-1 (AGO1) and direct the formation of silent chromatin domains. This mechanism is known as transcriptional gene silencing (TGS), and the development of TGS as a novel therapeutic modality would be applicable to chronic diseases where long-term, heritable silencing of target genes is warranted. Here we discuss how small RNAs can be used to direct TGS in mammalian cells.

Key words: RNAi, siRNA, microRNA, TGS, chromatin, epigenetic, methylation, transcription.

1. Introduction

1.1. RNAi and Transcriptional Gene Silencing

RNA interference (RNAi) is triggered by small RNAs that silence gene expression at the post-transcriptional level (1). Over the past few years, it has become increasingly apparent that components of the RNAi machinery also function in the nucleus of mammalian cells. Small interfering RNAs (siRNAs) and microRNAs (miRNAs) 22 nucleotides in length initiate sequence-specific gene silencing by loading into effector Argonaute (AGO) proteins (1), and when targeted to promoter regions, small RNAs induce transcriptional gene silencing (TGS) through Argonaute-1 (AGO1) and histone methylation (2). Marks of facultative heterochromatin, histone H3 lysine 9 dimethylation (H3K9me2) and histone H3 lysine 27 trimethylation (H3K27me3), are enriched at

John F. Reidhaar-Olson and Cristina M. Rondinone (eds.), *Therapeutic Applications of RNAi: Methods and Protocols, vol. 555*
© Humana Press, a part of Springer Science+Business Media, LLC 2009
DOI 10.1007/978-1-60327-295-7_9 Springerprotocols.com

promoter regions targeted by small RNAs *(2)* and can be measured using chromatin immunoprecipitation (ChIP) assays. In this chapter, we describe how to induce TGS in mammalian cells through the effective design of promoter-targeting siRNAs and how to assess the efficacy of silencing by examining chromatin marks at the silenced promoter, as well as using nuclear run-on assays.

1.2. Potential Therapeutic Applications of TGS

The use of TGS as a therapeutic modality would be applicable for targeting chronic diseases such as human immunodeficiency virus (HIV) infection and consequent acquired immune deficiency syndrome (AIDS) *(1)*. Targeting the long-terminal repeat (LTR) of HIV-1 with siRNAs has been shown to induce TGS *(3)*. Similar to TGS at endogenous promoters, the HIV-1 LTR exhibits H3K27me3 when targeted by siRNAs in the U3 region of the LTR. Moreover, the chemokine (C-C motif) receptor 5 (CCR5), which serves as a co-receptor for HIV-1, is amenable to TGS using siRNAs *(2)*. Knockdown of CCR5 is therapeutically relevant, since the loss of CCR5 would inhibit R5-tropic HIV-1 from gaining entry into its target cells. The development of therapeutic applications of TGS are still preliminary, however, and require further testing in primary cells before additional progress can be made.

2. Materials

2.1. Design of Promoter-Targeting siRNAs

1. http://www.invitrogen.com/rnai/
2. http://genome.ucsc.edu/

2.2. Assessing Histone Methylation Using ChIP

1. Synthetic siRNAs (Invitrogen).
2. Lipofectamine 2000 (Invitrogen).
3. 1% formaldehyde.
4. 0.125 M glycine.
5. 1X PBS.
6. PMSF.
7. ChIP lysis buffer, 50 mM HEPES at pH 7.5, 140 mM NaCl, 10% Triton X-100, 0.1% Sodium Deoxycholate (NaD), 1/1000 PMSF.
8. Bioruptor sonicator with refrigerated water bath and a rotating 1.5 mL microtube unit (Diagenode).
9. 50% Protein G Agarose slurry (Upstate).
10. Anti-AGO1 (Upstate: 07-599), Anti-H3K9me2 (Upstate: 07-441), and Anti-H3K27me3 (Upstate: 07-449).

11. ChIP lysis buffer high salt, 50 mM HEPES at pH 7.5, 500 mM NaCl, 1% Triton X-100, 0.1% NaD, 1/1000 PMSF.

12. ChIP wash buffer, 10 mM Tris-HCl at pH 8.0, 250 mM LiCl, 0.5% NP-40, 0.5% NaD, 1 mM EDTA.

13. Elution buffer, 50 mM Tris-HCl at pH 8.0, 1% SDS, 10 mM EDTA.

14. RNase A (10 mg/mL).

15. 5 M NaCl.

16. 0.5 M EDTA.

17. 1 M Tris-HCl at pH 6.5.

18. 10 mg/mL Proteinase K.

19. Phenol/chloroform.

20. iQ SYBR Green Supermix (Bio-Rad).

2.3. Measuring Transcriptional Silencing Using Nuclear Run-On

1. Synthetic siRNAs (Invitrogen).

2. Lipofectamine 2000 (Invitrogen).

3. 1X PBS.

4. 0.5% NP-40 lysis buffer, 10 mM Tris-HCl at pH 7.4, 10 mM NaCl, 3 mM $MgCl_2$.

5. Reaction buffer, 10 mM Tris-HCl at pH 8.0, 5 mM $MgCl_2$, 0.3 mM KCl.

6. 2.5 mM NTP + Biotin-16-UTP mix (Roche).

7. RNA STAT-60 Reagent (Tel-Test).

8. Streptavidin beads (Active Motif).

9. 2X SSC, 15% formamide.

10. 2X SSC.

11. 10 mM EDTA at pH 8.2.

3. Methods

3.1. Design of Promoter-Targeting siRNAs

Use an online siRNA design program, such as Invitrogen's RNAi designer, to enter the promoter sequence of a given gene of interest. Promoter sequences for human genes can be obtained from the UCSC Genome Brower. Select promoter sequences −1 to −200 bp upstream of the known gene transcription start site and enter them into the online siRNA design program. Choose up to four siRNAs that span the promoter region to test their efficacy in triggering TGS of the target gene. The efficacy of siRNA-mediated TGS may be enhanced with increasing proximity to the transcription start site, and several TGS studies *(2–6)* have shown targeting of promoter sequences −1 to −200 bp upstream of transcription start sites (**Table 9.1**). Important considerations for siRNA design include proper antisense or guide strand incorporation in AGO, since recent findings have shown that the antisense

Table 9. 1
Designing promoter-targeting siRNAs

TGS target genes	siRNA target sites
EF1A	−106 to −86
CDH1	−181 to −161
PR	−26 to −7
CCR5	−183 to −163
uPA	−131 to −111

strand alone is sufficient to induce TGS and heterochromatin formation, whereas the sense strand has no effect *(3)*. Furthermore, low levels of sense transcription across the targeted promoter region may be required for TGS to occur at a given gene promoter *(7)*.

3.2. Assessing Histone Methylation Using ChIP

Using siRNAs to target a promoter region induces epigenetic modifications that silence gene expression at the chromatin level. Histone methylation marks H3K9me2 and H3K27me3 are enriched in promoter regions targeted by siRNAs. Two studies have demonstrated that the CDH1 *(4)* and CCR5 *(2)* promoters exhibit increased levels of H3K9me2 when targeted with siRNAs. Additionally, the siRNA-targeted EF1A promoter exhibits both H3K9me2 and H3K27me3 during TGS *(3)*, suggesting that siRNAs might regulate the "histone code" through the recruitment of histone methyltransferases such as the Polycomb group component EZH2 *(2)*. Here we outline how to assess chromatin marks at siRNA-targeted promoters using ChIP, as previously described *(3)*.

1. Transfect cells with control or promoter-specific siRNAs at 50 nM final concentration using Lipofectamine 2000 (*see* **Note 1**).
2. 24–48 h following transfection, crosslink 2×10^7 cells with 1% formaldehyde for 10 min at room temperature (*see* **Note 2**).
3. Stop crosslinking by adding glycine at a final concentration of 0.125 M for 10 min at room temperature.
4. Wash cells twice with cold 1X PBS + 1/1000 PMSF.
5. Harvest cells in 1 mL of ChIP lysis buffer.
6. Incubate for 10 min on ice.
7. Centrifuge samples at $2,000 \times g$ for 5 min at 4°C and remove supernatant.
8. Resuspend pellet in 1 mL of ChIP lysis buffer and incubate on ice for 10 min.

9. Sonicate lysates using a Bioruptor sonicator with refrigerated water bath and a rotating 1.5-mL microtube unit for five cycles of 30 s ON and 30 s OFF at HIGH setting (*see* **Note 3**).

10. Centrifuge sonicated samples at $14,000 \times g$ for 10 min at 4°C.

11. Remove supernatant and preclear with 60 μL of 50% Protein G Agarose slurry for 1 h at 4°C with rotation.

12. Divide samples and perform immunoprecipitations using 1–5 μg of Anti-AGO1, Anti-H3K9me2, and/or Anti-H3K27me3 overnight at 4°C with rotation, along with no antibody controls.

13. Add 60 μL of 50% Protein G Agarose slurry to each sample for 1 h at 4°C with rotation.

14. Centrifuge at $1,200 \times g$ for 3 min at 4°C.

15. Save no antibody control supernatants and use as input control.

16. Perform two washes with 1 mL of ChIP lysis buffer, two washes with 1 mL of ChIP lysis buffer high salt, followed by two washes with 1 mL of ChIP wash buffer.

17. For each wash, incubate samples for 3 min at room temperature on a rotating platform, followed by centrifugation at $1,200 \times g$ for 3 min at room temperature.

18. Add 100 μL of elution buffer for 10 min at 65°C, followed by centrifugation at $14,000 \times g$ for 3 min at room temperature (repeat Step 18).

19. Reverse crosslinks in eluted samples by adding 1 μL of RNase A and 20 μL of 5 M NaCl and incubate for 4–6 h at 65°C.

20. Add 10 μL of 0.5 M EDTA, 20 μL of 1 M Tris-HCl at pH 6.5, and 2 μL of Proteinase K and incubate for 1 h at 45°C.

21. Recover DNA by phenol/chloroform extraction.

22. Use promoter-specific primers and iQ SYBR Green Supermix to assess the enrichment of chromatin marks using quantitative PCR (qPCR) (*see* **Note 4**).

3.3. Measuring Transcriptional Silencing Using Nuclear Run-On

Here we outline how to measure the extent of transcriptional silencing using nuclear run-on assays, as previously described *(8)*.

1. Transfect cells with control or promoter-specific siRNAs at 50 nM final concentration using Lipofectamine 2000.

2. 24–48 h following transfection, wash 2×10^7 cells twice with cold 1X PBS.

3. Harvest cells and lyse on ice in 0.5% NP-40 lysis buffer.

4. Centrifuge at $500 \times g$ for 10 min.

5. Remove supernatants.

6. Incubate nuclei in reaction buffer and 2.5 mM NTP + Biotin-16-UTP mix for 45 min at 30°C.

7. Stop transcription reaction by adding RNA-STAT-60 and recover RNA.

8. Isolate biotinylated nascent RNA transcripts by incubation with streptavidin beads for 2 h at room temperature on a rocking platform.

9. Centrifuge at $3{,}000 \times g$ for 3 min to collect beads.

10. Wash once with 2X SSC, 15% formamide for 10 min on a rocking platform.

11. Wash twice with 2X SSC for 5 min on a rocking platform.

12. Elute biotinylated RNA from streptavidin beads in H_2O or 10 mM EDTA at pH 8.2 by incubating at 90°C for 10 min.

13. Use mRNA-specific primers to analyze nascent RNA transcript levels using qRT-PCR.

4. Notes

1. Analogous to traditional RNAi experiments, it is critical to optimize transfection conditions for a given cell type. While the use of Lipofectamine 2000 has been described, certain cell types may not exhibit a high level of transfection efficiency using this reagent.

2. Some cell types may exhibit slower doubling times and may require longer periods of time to establish TGS following siRNA transfection. In these instances, waiting 48–72 h after siRNA transfection may provide a better window to assess silent chromatin marks using ChIP.

3. For ChIP experiments, optimization of chromatin shearing conditions is critical and must be determined separately for each cell type.

4. Some genes may not be amenable to siRNA-mediated TGS. Low levels of sense transcription across the target promoter region may be necessary for the siRNA antisense or guide strand to recognize and target. Additionally, highly euchromatic genes may not be susceptible to siRNA-directed heterochromatin formation.

Acknowledgments

This work was supported by a City of Hope Graduate School Internal Fellowship to D.H.K. and National Institutes of Health grants R37-AI29329 and R01-AI42552 to J.J.R.

References

1. Kim, D. H. and Rossi, J. J. (2007) Strategies for silencing human disease using RNA interference. *Nat. Rev. Genet.* **8**, 173–184.

2. Kim, D. H., Villeneuve, L. M., Morris, K. V., and Rossi, J. J. (2006) Argonaute-1 directs siRNA-mediated transcriptional gene silencing in human cells. *Nat. Struct. Mol. Biol.* **13**, 793–797.

3. Weinberg, M. S., Villeneuve, L. M., Ehsani, A., Amarzguioui, M., Aagaard, L., Chen, Z. X., Riggs, A. D., Rossi, J. J., and Morris, K. V. (2006) The antisense strand of small interfering RNAs directs histone methylation and transcriptional gene silencing in human cells. *RNA* **12**, 256–262.

4. Ting, A. H., Schuebel, K. E., Herman, J. G., and Baylin, S. B. (2005) Short double-stranded RNA induces transcriptional gene silencing in human cancer cells in the absence of DNA methylation. *Nat. Genet.* **37**, 906–910.

5. Janowski, B. A., Huffman, K. E., Schwartz, J. C., Ram, R., Hardy, D., Shames, D. S., Minna, J. D., and Corey, D. R. (2005) Inhibiting gene expression at transcription start sites in chromosomal DNA with antigene RNAs. *Nat. Chem. Biol.* **1**, 216–222.

6. Pulukuri, S. M. and Rao, J. S. (2007) Small interfering RNA directed reversal of urokinase plasminogen activator demethylation inhibits prostate tumor growth and metastasis. *Cancer Res.* **67**, 6637–6646.

7. Han, J., Kim, D., and Morris, K. V. (2007) Promoter-associated RNA is required for RNA-directed transcriptional gene silencing in human cells. *Proc. Natl. Acad. Sci. USA* **104**, 12422–12427.

8. Zhang, M. X., et al. (2005) Regulation of endothelial nitric oxide synthase by small RNA. *Proc. Natl. Acad. Sci. USA* **102**, 16967–16972.

Chapter 10

Alternative Splicing as a Therapeutic Target for Human Diseases

Kenneth J. Dery, Veronica Gusti, Shikha Gaur, John E. Shively, Yun Yen, and Rajesh K. Gaur

Abstract

The majority of eukaryotic genes undergo alternative splicing, an evolutionarily conserved phenomenon, to generate functionally diverse protein isoforms from a single transcript. The fact that defective pre-mRNA splicing can generate non-functional and often toxic proteins with catastrophic effects, accurate removal of introns and joining of exons is vital for cell homeostasis. Thus, molecular tools that could either silence a disease-causing gene or regulate its expression in *trans* will find many therapeutic applications. Here we present two RNA-based approaches, namely RNAi and theophylline-responsive riboswitch that can regulate gene expression by loss-of-function and modulation of splicing, respectively. These strategies are likely to continue to play an integral role in studying gene function and drug discovery.

Key words: Alternative splicing, spliceosome, RNAi, theophylline-responsive riboswitch.

1. Introduction

The diversity of the human proteome is attributable in large part to the complexity and processing of pre-mRNAs via a process known as RNA splicing. This biochemical pathway leads to the assembly onto pre-mRNA of the mature spliceosome, which catalyzes the excision of non-coding sequences from precursor mRNA. A pre-mRNA is defined by four *cis*-acting elements: the 5′ and 3′ splice sites (ss), the branchpoint sequence, and the poly(Y)-tract. These *cis*-acting elements direct the assembly of the basal splicing machinery that leads to the removal of introns and joining of exons. In general, pre-mRNAs that undergo alternative

John F. Reidhaar-Olson and Cristina M. Rondinone (eds.), *Therapeutic Applications of RNAi: Methods and Protocols, vol. 555*
© Humana Press, a part of Springer Science+Business Media, LLC 2009
DOI 10.1007/978-1-60327-295-7_10 Springerprotocols.com

splicing harbor suboptimal splicing signals and often require the help of auxiliary regulatory elements known as splicing enhancers and silencers for splice site pairing. These elements provide the binding surface for splicing regulatory proteins that communicate with the basic splicing machinery to define splice site choice (for review *see* Ref. *1–3*).

Alternative splicing of pre-mRNA is now considered to be the most important source of protein diversity in vertebrates. It is estimated that more than 60% of human genes generate transcripts that are alternatively spliced (*4, 5*) and 70–90% of alternative splicing events affect coding capacity of genes (*6*). Importantly, deregulated splice variant expression has been identified as the cause of a number of genetic disorders, and certain forms of cancer have been linked to unbalanced isoform expression from genes involved in cell cycle regulation or apoptosis (*7, 8*). Given the critical role of alternative splicing in a variety of cellular processes (*9*), strategies that could influence pre-mRNA splicing decisions will have far-reaching effects in biotechnology and medicine. This chapter focuses on the design and construction of two RNA-based molecular approaches, namely RNA interference (RNAi) and theophylline riboswitch, with the ultimate aim of regulating gene expression. Whereas RNAi exerts its effect in *trans* to downregulating the expression of a target gene, theophylline-responsive riboswitch controls gene expression by modulating pre-mRNA splicing.

The use of RNAi as a gene-silencer strategy represents a powerful tool in the field of small-molecule nucleic acid-based therapeutics. Described first in *Caenorhabditis elegans*, this biological phenomenon has subsequently been studied in a wide range of organisms (for recent review, *see* (*10*)). RNAi involves double-stranded RNA molecules of approximately 20–25 nucleotides termed short interfering RNAs (siRNAs) that are processed by the endogenous RNAse III family member Dicer and are incorporated into an RNA-induced silencing complex (RISC) in a process that prevents the expression of a particular gene (*11*). The relative ease with which a siRNA can be designed and synthesized, its specificity and potency and, most importantly, the ability to preferentially suppress the expression of mutant alleles makes this approach highly appealing. Moreover, RNAi not only has the potential to be an effective therapeutic tool, but also enables the identification of genes that regulate alternative splicing (*7, 12*). However, RNAi is faced with limitations when compared with other approaches such as the theophylline-responsive riboswitch and antisense oligonucleotides, which unlike RNAi, can modulate mRNA isoform levels (*13–16*).

Riboswitches (*17, 18*) are natural RNA aptamers that regulate gene expression by binding to small-molecule ligands. As RNA structures (*14, 19*) are known to influence splice site choice, we hypothesized that sequestering of splicing regulatory elements within RNA secondary structure could influence splice site choice. By exploiting the ligand-induced conformational rearrangement property of theophylline riboswitch we have demonstrated the control of alternative splicing both in vitro and in vivo (*13, 14, 20*). This novel technology represents the possibility for controlling splicing of a *trans* gene in a gene therapy setting where the target gene expression could be controlled in a ligand-dependent manner.

2. Materials

2.1. RNA Interference

1. All buffers and solutions were made using filtered deionized, 18 MΩ water purified by a Barnstead MP-3A Megapure system.
2. Fetal bovine serum (Irvine Scientific, CA, USA).
3. MCF7 cells: This is a malignant mammary epithelial cell line (ATCC, Manassas, VA, http://www.atcc.org/).
4. Minimum essential medium (MEM): 2 mM L-glutamine and Earle's BSS adjusted to contain 1.5 g/l sodium bicarbonate, 0.1 mM non-essential amino acids, and 1 mM sodium pyruvate and supplemented with 0.01 mg/ml bovine insulin; and 10% fetal bovine serum (ATCC, Manassas, VA, http://www.atcc.org/).
5. Lipofectamine 2000 transfection reagent (Invitrogen, CA, USA).
6. Opti-MEM (Invitrogen, CA, USA).
7. siRNAs: We synthesized the following RNAi oligos against human CEA: sense 5′-CUGGCCAGUUCCGGGUAUA-3′ and antisense 5′-UAUACCCGGAACUGGCCAG-3′ (nucleotides 404–422, numbering from the initial start codon); sense 5′-CGGGACCUAUGCCUGUUUU-3′ and antisense 5′-AAAACAGGCAUAGGUCCCG-3′ (nucleotides 1950–1968, numbering from the initial start codon), (Qiagen, CA, USA). A scrambled control siRNA was synthesized at the City of Hope DNA/RNA core facility and is randomized with respect to its nucleotide distribution. The siRNAs were diluted in water to a stock concentration of 20 μM.

2.2. Western Blot

1. ECL plus Western Blotting detection reagents (Amersham Biosciences, Buckinghamshire, England).
2. Blocking buffer: 150 mM NaCl, 5 mM EDTA, 50 mM Tris-HCl, pH 7.5, 0.05% Triton X-100, 0.25% gelatin. Blocking

buffer can be made as a 10x stock and stored at room temperature without gelatin. Store the 1x dilution at 4°C.

3. Lysis buffer: 10 mM Tris-HCl, pH 8, 140 mM NaCl, 0.025% sodium azide, 1% Triton X-100, 1 mM EDTA, 1 mM PMSF, and 1 mM sodium vanadate.

4. 2x Gel loading buffer: 0.125 M Tris-HCl, pH 6.8, 4% SDS, 20% glycerol, 10% 2-mercaptoethanol, 0.005% bromophenol blue.

5. Anti-β-actin Goat Polyclonal IgG (Santa Cruz Biotechnology, Inc, CA, USA). Dilution for immunoblots, 1:2000.

6. Donkey α-Goat IgG-HRP (Santa Cruz Biotechnology, Inc, CA, USA). Dilution for immunoblots, 1:5000.

7. Goat α-Mouse IgG-HRP (Santa Cruz Biotechnology, Inc, CA, USA). Dilution for immunoblots, 1:5000.

8. α-CEA T84.66: This is a chimeric monoclonal antibody of high specificity for tumor-associated CEA (22). Dilution for immunoblots, 1:2000.

9. Immun-blot PVDF/Filter Paper (BioRad, CA, USA).

2.3. Theophylline-Responsive Riboswitch

1. α-[^{32}P] UTP: 3000 Ci/mmol, 10 μCi/μl (Perkin Elmer, MA, USA).

2. 8% Denaturing polyacrylamide gel: From the Sequa-gel sequencing system kit (*see* **Section 2.3**, Step 17), mix 6.4 ml of Sequa-gel concentrate, 11.6 ml of Sequa-gel diluent, and 2 ml of Sequa-gel buffer. To this solution, add 200 μl of 10% ammonium persulfate and 20 μl of TEMED. Swirl gently to mix and cast the gel.

3. BC300: 20 mM HEPES, pH 8.0, 20% glycerol, 300 mM KCl, 0.2 mM EDTA.

4. Dithiothreitol (DTT): To prepare a 0.1 M stock, add 0.154 g of DTT in 10 ml RNAse-free water.

5. Dulbecco's Modified Eagle medium supplemented with L-glutamine and 10% fetal bovine serum (Omega Scientific, CA, USA).

6. Electrophoresis sample buffer: Add 2.4 ml 1 M Tris-HCl, pH 6.8, 3 ml 20% SDS, 3 ml 100% glycerol, 1.6 ml 2-mercaptoethanol, 0.006 g bromophenol blue. Store at 4°C.

7. Exon junction primers: The following primers were used to assess theophylline-mediated modulation of alternative splicing in vivo: #43573, 5′-G GGCCAGCTGTTGGGGTCGA-3′ and #43754, 5′-GGGCCAGCTGTTGGGCTCGC-3′ as forward primers and oligonucleotide #39368, 5′-TAGA GGATCCCCACTGGAAAGACCG-3′ as reverse primer.

8. 5x First-strand buffer for cDNA synthesis: 250 mM Tris-HCl, pH 8.3, 375 mM KCl, 15 mM MgCl$_2$.

9. Glycogen: Stock solution is 20 mg/ml (Roche Applied Science, Mannheim, Germany). Dilute to 5 mg/ml and store in small aliquots at −20°C.

10. HeLa nuclear extract for in vitro splicing: HeLa nuclear extract was prepared from 10 l HeLa-S3 cells (National Cell Culture Center, MN, USA) as described previously (21). HeLa nuclear extracts can also be purchased from 4C Biotech (Seneffe, Belgium).

11. Lipofectamine 2000 (Invitrogen, CA, USA).

12. Molecular Dynamics PhosphorImager and the ImageJ software (http://rsb.info.nih.gov/ij/) were used to quantitate the products of the in vitro RNA splicing reaction and RNAi experiments.

13. M-MLV reverse transcriptase (Invitrogen, CA, USA).

14. Opti-MEM Reduced Serum Medium (Invitrogen, CA, USA).

15. Proteinase K (Roche Applied Science, Mannheim, Germany).

16. RNasin: 10 units/10 μl reaction (Promega, WI, USA).

17. Sequa-gel sequencing system kit: Contains Sequa-gel concentrate, Sequa-gel diluent, and Sequa-gel buffer (National Diagnostics, Georgia USA).

18. Splicing loading buffer: Mix 2 mg of xylene cyanol and 2 mg of bromophenol blue in 10 ml of deionized formamide.

19. 7mG(ppp)G RNA cap structure analog: Dilute the stock in 131 μl RNase-free water to give a working concentration of 10 mM. For transcription reactions, use a 5:1 ratio of 7mG(ppp)G RNA cap structure analog to GTP.

20. Splicing stop buffer: To stop in vitro splicing reaction, add 50 μl 2x PK buffer, 2.5 μl 10 mg/ml Proteinase K, 2 μl of 5 mg/ml glycogen, and RNase-free water to a total volume of 87.5 μl.

21. 2x PK buffer: 200 mM Tris-HCl, pH 7.5, 25 mM EDTA, pH 8.0, 300 mM NaCl, 2% SDS.

22. 10x RNA polymerase reaction buffer: 400 mM Tris-HCl, pH 7.9, 60 mM $MgCl_2$, 20 mM spermadine, 100 mM DTT.

23. 10x transcription master mix: Combine 3.1 μl each of 4 mM GTP, 20 mM ATP, 20 mM CTP, 4 mM UTP, and 10 μl of 0.1 M DTT. The mixture is gently agitated and briefly micro-centrifuged to bring the mixture to homogeneity. Dispense 2.24 μl per transcription reaction. Store mix @ −20°C with limited freeze–thaw cycles.

24. 25 mM Theophylline: Add 0.09 g theophylline (Sigma-Aldrich, MO, USA) to 20 ml of RNase-free water. For a final concentration of 2 mM, add 2 μl to a 25 μl splicing reaction.

25. T7 RNA polymerase: 10 units/10 μl reaction (New England Biolabs, MA, USA).

26. 13% denaturing polyacrylamide gel: From the Sequa-gel sequencing system kit (*see* **Section 2.3**, Step 17), mix 10.4 ml of Sequa-gel concentrate, 7.6 ml of Sequa-gel diluent, and 2 ml of Sequa-gel buffer. To this solution, add 200 μl of 10% ammonium persulfate and 20 μl of TEMED. Swirl gently to mix and cast the gel.

3. Methods

3.1. siRNA-Mediated Degradation of CEA mRNA

The siRNA inhibits gene expression by degrading the corresponding endogenous mRNA. RNAi has been successfully used in establishing the role of splicing regulatory proteins in human cancer (*22–25*). In addition, siRNA-mediated downregulation of Bcl-xL isoform in TRAIL-resistant cells has been shown to inhibit cell proliferation and sensitize TRAIL-induced apoptosis in human cancer cells with both acquired and intrinsic TRAIL resistance (*26*). However, the design of siRNA that targets with high efficiency and specificity is critical to successful gene silencing. Here we show that RNAi can be used to downregulate carcinoembryonic antigen (CEA), a glycoprotein that is overexpressed in a variety of cancers (*27, 28*).

3.1.1. siRNA Design

Prior to finding an RNAi target on the gene of interest, the mRNA sequence or sequence accession number should be retrieved from Refseq at NCBI. The Entrez query tool can be located at http://www.ncbi.nlm.nih.gov/entrez/query.fcgi?db=Nucleotide. On the "Limits" tab, pull down the "Only from" menu and choose Refseq. This will restrict the query to the RefSeq database only. Other potential search routes are: (1) Search LocusLink by gene name or symbol at http://www.ncbi.nlm.nih.gov/LocusLink/ and (2) Search Entrez Gene at http://www.ncbi.nlm.nih.gov/entrez/query.fcgi?db=gene (*see* (*29*), **Table 1** for a list of other Web servers for siRNA design). To prevent off-target effects, check the gene of interest for significant homology to other genes or sequences in the genome using the NCBI Blast tool – Nucleotide–nucleotide BLAST (blastn), or Blat tool on UCSC Genome Website http://genome. ucsc.edu/cgi-bin/hgBlat, or Ensembl Blast at http://www.ensembl.org/Multi/blastview.

Many different algorithms and programs that depend on sequence characteristics or target mRNA secondary structures are available now for the design of siRNAs for post-transcriptional gene silencing (*29–32*). The whole length of an mRNA sequence can be targeted by RNAi though it is especially important to be cognizant that not all siRNAs target the mRNA with equal efficiency. It is crucial to perform a homology search for candidate siRNA targets against other mRNA sequences of the genome. In general, a siRNA should not have significant homology with unrelated sequences especially at its 3' end. If a siRNA design program does not provide the function for homology sequence, a manual BLAST search is mandatory for avoiding off-target effects. The main challenge for developing siRNA in vivo is delivering duplex RNA intact to a target tissue. Many of these pharmacokinetic obstacles also confront antisense oligonucleotides. However, a crucial difference between the two is that antisense oligonucleotides comprise just one nucleic acid strand, whereas siRNA is made up of two strands. On a practical level, the mass of a synthetic duplex RNA is twice that of a traditional antisense oligonucleotide and can result in increased costs. The higher molecular weight of siRNA can also make uptake by cells more difficult. Furthermore, the presence of a second nucleic acid strand increases the potential for off-target effects. For effective delivery, high duplex stability is needed because duplex RNA must remain hybridized until it enters the RISC complex. The most effective siRNA is identified experimentally with rules that govern effective siRNA design dependent on the particular application. Here we present some sequence-characteristic-based guidelines to design high-efficiency siRNAs (33, 34):

1. Regions located 50–100 nt downstream of the start codon (ATG) make good targets for RNAi.
2. Sequence motifs that have $AA(N_{19})TT$, $NA(N_{21})$, or $NAR(N_{17})YNN$ where N is any nucleotide, R is purine (A, G), and Y is pyrimidine (C, U) make good targets for RNAi.
3. Avoid introns. As siRNAs are processed only in the cytoplasm and not within the nucleus, post-transcriptional processing would prevent targeting of the siRNA.
4. Sequences with high G + C content (more than 50%) should be avoided. The GC content of the sense strand should be between 30 and 50%.
5. Stretches of polynucleotide repeats should be avoided.

3.1.2. RNAi Against CEA

To investigate the relationship between overexpression of CEA and breast cancer progression, we employed two chemically synthesized siRNAs (siRNAs #497 and #2043) to downregulate CEA mRNA (*see* **Note 1**). To monitor the cellular uptake, a scrambled control siRNA (Sc) bearing fluorescein at the 5' end of the antisense strand was also synthesized. MCF7 cells that are known to overexpress CEA were transiently transfected with different siRNAs (*see* **Section 3.1.3** for details on animal cell transfection, *see* **Note 2**). After indicated time (we recommend performing a time course), cells were harvested and protein lysates were resolved on a 4–20% SDS-polyacrylamide gradient gel followed by transfer to a polyvinylidene fluoride (PVDF) membrane. Transferred proteins were probed with either anti-CEA or anti-β-actin antibody. Our results demonstrate that compared to Sc control, siRNA #497 significantly downregulated CEA protein (**Fig. 10.1A**) (*see* **Note 3**). By contrast, siRNA #2043 showed moderate downregulation of CEA protein (54%). This observed difference could be due to the accessibility of the target region and not unequal loading of sample as evidenced by the levels of probes to β-actin protein (**Fig. 10.1A**). We note that high dosages of siRNA #497 are more effective at gene silencing; suggesting that nuclease digestion likely limits the efficacy at the lower concentration (**Fig. 10.1A**, compare lane 4 with lane 5). A time course presented in **Fig. 10.1B** demonstrates that knockdown of CEA protein occurs as early as 24 h and is effective at least up to 72 h post-treatment. The slight increase in protein levels at 72 h is apparently the effect of siRNA depletion (**Fig. 10.1B**, compare lane 3 with lane 4).

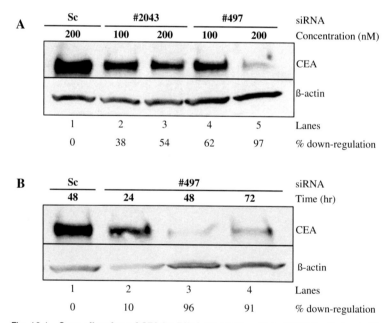

Fig. 10.1. **Gene silencing of CEA by RNAi. (A)** Western blot of MCF7 cells treated for 48 h with siRNAs #2043, #497, and a scrambled (Sc) control at a concentration of 100 nM (lanes 2 and 4) or 200 nM (lanes 1, 3, and 5). **(B)** MCF7 cells treated after 24, 48, and 72 h with 200 nM siRNAs #497 or 200 nM Sc as indicated. Upper panels in both A and B show a Western blot probed with α-CEA T84.66 while the lower panels were treated with anti-β-actin antibody. The percentage of downregulation is indicated below each panel and calculated by comparing CEA downregulation to β-Actin and normalizing to the scrambled control.

3.1.3. Transient Transfection of MCF7 Cells

1. In 2 ml MEM supplemented with 10% fetal bovine serum, seed the following number of cells in each well of a six-well plate 24 h prior to transfection: 6×10^5 cells (70% confluency for harvest time at 24 h), 5×10^5 cells (60% confluency for harvest time at 48 h), and 4×10^5 cells (50% confluency for harvest time at 72 h).

2. Cells should be maintained under standard incubation conditions (humidified atmosphere, 5% CO_2, 37°C) (*see* **Note 4**).

3. Next day, prepare two mixtures in separate eppendorf tubes (A and B). Tube A contains either 100 or 200 nM siRNA diluted with Opti-MEM (*see* **Note 5**) to a final volume of 250 µl. Tube B contains 2 µl Lipofectamine 2000 (*see* **Note 6**) diluted with Opti-MEM to a final volume of 250 µl. A master mix is recommended if all the wells to be transfected contain the same siRNA.

4. After an incubation period of 5 min, both tubes A and B are combined and incubated for 20 min to allow complex formation between siRNA and the Lipofectamine 2000.

5. Prior to the completion of the 20-min incubation period, aspirate the media from cells and replace with 2 ml fresh media to each well.

6. Add RNAi–Lipofectamine 2000 complex (500 µl per well) dropwise. Distribute around the well and gently swirl the plate.

7. Cells should be maintained under standard incubation conditions (humidified atmosphere, 5% CO_2, 37°C) for an additional 24–72 h depending on the harvest time.

8. Assay the cells using the appropriate post-treatment protocol (see **Section 3.1.4** for Western Blot Protocol and **Section 3.2.7** for RT-PCR analysis; *see* **Note 7**).

3.1.4. Western Blot Analysis

1. After the indicated time, cells are harvested and the lysate is prepared by adding 150–200 µl of lysis buffer.

2. The cells are homogenized by the passage (approximately 10 times) through a 1-ml syringe with a 22-gauge needle.

3. Next, spin the homogenized cell lysate @ $10,000 \times g$ for 20 min at 4°C. Remove the supernatant and to the cell pellet add 2x Gel loading buffer.

4. The protein lysates were resolved on a 4–20% SDS-polyacrylamide gradient gel essentially as described earlier (*35*). Next, the proteins were transferred to an Immun-Blot PVDF membrane and incubated in blocking buffer for 60 min at room temperature on a shaking platform (with a change of buffer after 30 min).

5. The primary antibodies, either α-CEA T84.66 or anti-β-actin Goat Polyclonal IgG, are next added to the blocking

buffer and incubated in the presence of the membrane overnight at room temperature on a shaking platform (*see* **Note 8**).

6. Next day, the membrane is washed three times (20 min each rinse) in blocking buffer followed by an additional incubation with the secondary antibody (either Goat α-Mouse IgG-HRP or Donkey α-Goat IgG-HRP) for 1 h in blocking buffer.

7. After washing, develop the membrane using Amersham ECL Plus Western Blotting detection reagents, following the manufacturer's recommendation (*see* **Note 9**).

3.2. Theophylline-Responsive Riboswitch

It is widely accepted that deregulated alternative splicing is the cause of a number of human diseases. It has been estimated that approximately 15% of all mutations that cause genetic diseases result in defective splicing of pre-mRNA: defects in splicing has been shown to be associated with genetic disorders such as β-thalassemia, cystic fibrosis, Duchenne muscular dystrophy, etc. In addition, certain forms of cancer have been linked to unbalanced isoform expression from genes involved in cell cycle regulation or angiogenesis (reviewed in Refs. (*12, 36*)). Given the significance of alternative splicing in generating protein diversity and its link to human diseases, mRNA splicing has become an important target of therapeutic intervention (*37*).

We have recently demonstrated that a theophylline-responsive riboswitch, an RNA aptamer that binds to theophylline with high affinity and specificity, can modulate alternative splicing both in vitro and in cultured cells (*13, 14, 20*). This approach can be modified to regulate the expression of virtually any *trans*-gene by controlling its splicing. Here we describe the experimental details for controlling pre-mRNA splicing with the theophylline-responsive riboswitch.

3.2.1. Design of Theophylline-Responsive Riboswitch

Theophylline riboswitch is an RNA aptamer consisting of a 15-nt core sequence, which provides affinity and specificity for RNA–theophylline interaction (**Fig. 10.2A**). Remarkably, caffeine that differs from theophylline by only a methyl group at the N-7 position in the imidazole ring binds to theophylline aptamer with 10,000-fold less affinity (*38*). Theophylline aptamer is unstructured; however, ligand-induced conformational rearrangement of RNA allows theophylline binding by formation of thermodynamically stable RNA–theophylline complex. Although theophylline riboswitch can be adapted to control pre-mRNA splicing by targeting a variety of *cis*-acting regulatory elements, here in this section we focus our attention at the branchpoint sequence. We constructed a model pre-mRNA comprising of three exons interrupted by two introns (ABT4M, **Fig. 10.2B**). Our data indicate that sequestering of intron 1 branchpoint by theophylline

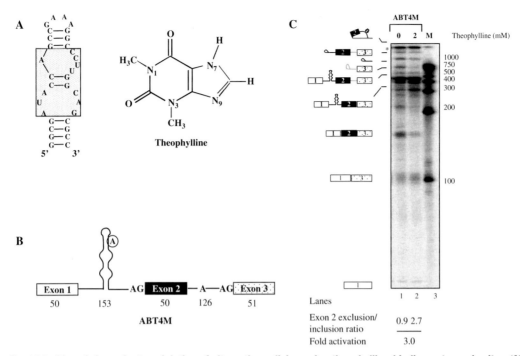

Fig. 10.2. **Ligand-dependent modulation of alternative splicing using theophylline-binding aptamer in vitro. (A)** Diagram of the theophylline RNA aptamer sequence (*left panel*) and the chemical structure of theophylline (*right panel*). The residues that are conserved for theophylline binding are enclosed in the rectangular box. **(B)** Schematic representation of a theophylline-binding aptamer sequestering the branchpoint of a model pre-mRNA. The encircled adenosine residue represents branch nucleotide. Open boxes represent exon sequences and horizontal lines between exons indicate introns. The numbers indicate size of the exon or intron. **(C)** In vitro splicing of ABT4M pre-mRNA. ^{32}P-labeled ABT4M pre-mRNA was subjected to in vitro splicing in the absence (lane 1) or presence of 2 mM theophylline (lane 2). The extracted RNAs were fractionated on a 13% polyacrylamide denaturing gel. Schematic representations of the various RNA species are indicated on the left. Molecular Dynamics PhosphorImager and the ImageJ software were used to quantitate the products of the in vitro and in vivo RNA splicing reactions (*see* **Fig. 10.3**).

allows intron 2 branchpoint to choose between the 5′ ss of exon 1 and 2 for the first step of splicing. Thus, depending upon which of the two 5′ ss is utilized determines the amount of exon 2 included/excluded mRNA (*see* **Sections 3.2.2** and **3.2.5**). Since theophylline aptamer's lower stem affects thermodynamic stability of RNA–theophylline complex, splicing can be fine-tuned by simply altering the size of the stem (*14, 19*) (*see* **Notes 10, 11**).

3.2.2. Theophylline-Responsive Riboswitch for the Modulation of Pre-mRNA Splicing In Vitro

^{32}P-labeled ABT4M pre-mRNA was transcribed (*see* **Section 3.2.3**) and incubated in HeLa nuclear extract in the absence or presence of theophylline (**Fig. 10.2C**). Splicing of ABT4M substrate gave rise to two spliced products, a slower migrating band that represents full-length mRNA and a faster moving band which represents exon 2-skipped mRNA generated as a result of alternative splicing. Notably, theophylline shifted the splicing of ABT4M pre-mRNA in favor of exon 2-excluded isoform by repressing the activation of intron 1 branchpoint.

3.2.3. In Vitro Transcription

T7 RNA polymerase transcription reactions were prepared in 12.5 μl volumes containing 1x RNA polymerase reaction buffer: 0.5 mM (ATP, CTP, GTP), 0.1 mM UTP, 10 mM DTT, 500 ng of linearized DNA template, 20 μCi α-[^{32}P] UTP (3000 Ci/mmol), 0.5 unit RNasin, and 0.5 units of T7 RNA polymerase. Assemble this reaction by preparing a 10x Transcription master mix (*see* Materials **Section 2.3**). To each pre-chilled sample eppendorf tube, the following is added in order:

1. 2.5 μl of RNase-free water
2. 2.2 μl of 10x Transcription master mix
3. 2.5 μl 7mG(ppp)G RNA cap structure analog
4. 1.3 μl RNA polymerase reaction buffer
5. 1 μl of DNA template
6. 0.5 μl of T7 RNA polymerase
7. 0.5 μl of RNasin
8. 2 μl α-[32P] UTP
9. Incubate at 37°C for 120 min.

Reactions are terminated by adding an equal volume of splicing loading buffer and heated at 65°C for 5 min. Briefly incubate the tube on ice and resolve by 8% denaturing PAGE gel electrophoresis.

3.2.4. In Vitro Splicing Assay

The extract preparation and in vitro splicing assay has been described previously (*21, 39*). Assemble the splicing reaction in the following order:

1. 1 μl BC300
2. 0.5 μl 160 mM MgCl$_2$
3. 1 μl of 10,000 c.p.m labeled RNA substrate
4. 1 μl 0.5 M Creatine phosphate
5. 0.5 μl 25 mM ATP
6. 0.25 μl 0.1 M DTT
7. 0.25 μl RNasin
8. 2 μl 25 mM Theophylline or RNase-free water in control splicing reaction
9. RNase-free water to a volume of 12.5 μl
10. 12.5 μl HeLa nuclear extract
11. Incubate at 30°C for 2 h and terminate by the addition of 87.5 μl splicing stop buffer.
12. Extract the RNA from the reaction mixtures by phenol-chloroform treatment followed by ethanol precipitation.
13. The RNA pellet is washed with 70% aqueous ethanol, dried, and dissolved in 10 μl loading buffer.
14. Splicing intermediates and products are analyzed by electrophoresis on a 13% denaturing polyacrylamide gel run at 50 W for 6 h.

3.2.5.
Theophylline-Dependent
Control of Alternative
Splicing In Vivo

To determine whether theophylline-induced sequestering of branchpoint can control splicing in living cells, we inserted the DNA that encodes ABT4M pre-mRNA into the mammalian expression vector pcDNA Myc/His yielding pcABT4M. A mutant (pcABT4Mmu) that does not bind to theophylline was also constructed. Next, HeLa cells were transfected transiently (*see* **Section 3.2.6**) with pcABT4M, pcABT4Mmu, or empty vector and treated with theophylline or buffer. After 24 h, cells were harvested, and total RNA was isolated. Reverse transcription followed by PCR shown in **Fig. 10.3A** demonstrates that theophylline can regulate the alternative splicing of ABT4M pre-mRNA (compare lane 4 with 5). In contrast, theophylline had

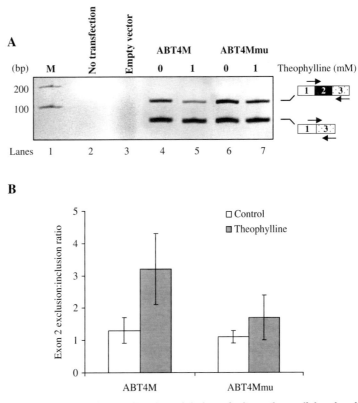

Fig. 10.3. **Theophylline-mediated modulation of alternative splicing in vivo.**
(A) HeLa cells were transiently transfected with empty vector, pcABT4M or pcABT4Mmu (containing mutations within core theophylline-binding aptamer). Cells were treated with buffer (lanes 4 and 6) or with 1 mM theophylline (lanes 5 and 7). The product of RT-PCR assays of total RNA isolated from each well was analyzed on a 2.5% agarose gel to estimate the effect of theophylline on the efficiency of exon 2 alternative splicing. The PCR-amplified bands corresponding to exon 2-included or -excluded mRNA is indicated. **(B)** The internal exon 2 exclusion: inclusion ratio for ABT4M or ABT4Mmu in the absence (*open box*) or presence of theophylline (*shaded box*) is shown. The data represent mean ± SEM. $P < 0.04$ versus mutant is significant.

a less significant effect on the splicing of ABT4Mmu pre-mRNA (**Fig. 10.3A**, compare lane 6 with 7; also see **Fig. 10.3B**), suggesting that binding of theophylline to its cognate RNA is necessary for controlling alternative splicing.

3.2.6. Transient Transfection of HeLa Cells

1. HeLa cells were grown in Dulbecco's modified Eagle medium.
2. Cells should be maintained under standard incubation conditions (humidified atmosphere, 5% CO_2, 37°C).
3. Seed 4.5×10^5 cells in a six-well plate a day prior to transfection.
4. Cells were grown to approximately 60–70% confluency and transfected with 1 μg of DNA by using Lipofectamine 2000 according to the manufacturer's protocol except that prior to the addition of DNA–Lipofectamine 2000 complex, the media was replaced by fresh media containing the indicated concentration of theophylline.
5. A freshly prepared solution of theophylline was used. In control samples, theophylline was replaced by buffer medium (*see* **Note 12**).

3.2.7. First-Strand cDNA Synthesis (RT-PCR)

1. In a sterile eppendorf tube we add 5 μg (total RNA), 500 ng oligo $(dT)_{12\text{-}18}$, 10 mM dNTP, and nuclease-free H_2O to final volume of 12 μl.
2. The mixture is heated for 5 min at 65°C. Spin briefly and place promptly on ice.
3. Add in the following order:
   1. 4 μl 5x First-Strand Buffer
   2. 2 μl 0.1 M DTT
   3. 1 μl RNasin
4. We then gently mix the contents of the tube and incubate for 2 min at 37°C.
5. Add 1 μl M-MLV Reverse Transcriptase (*see* **Note 13**).
6. Incubate at 37°C for 50 min.
7. Inactivate enzyme at 70°C for 15 min.
8. Store products at –20°C or proceed to amplification step using PCR.

4. Notes

1. Custom siRNA synthesis should be designed with high purity for animal studies and preclinical applications. Our siRNAs are obtained from Qiagen (Valencia, CA). We note that some companies (Santa Cruz Biotechnology) are now offering siRNA gene silencers targeted to all human genes. The RNAi molecules are provided as pools of three targets

specific to downregulate the expression of the gene of interest.

2. Always keep RNA solutions on ice to prevent degradation. Store unused siRNA duplex solution frozen at −20°C. The siRNA duplex solution can be frozen and thawed many times.

3. Although 200 nM of siRNA #497 was required to knock-down CEA, we recommend titration of siRNA to identify the optimum dose.

4. Healthy and sub-confluent cells are required for successful transfection experiments. Adjust cell and reagent amounts proportionately for wells or amount of DNA transfected. The objective is to have 90–100% confluency of cells at the time of harvest. The number of cells seeded varies depending on cell line.

5. Opti-MEM is a versatile, chemically defined medium, formulated to significantly reduce the amount of serum required for cultivating mammalian cells in vitro.

6. Lipofectamine 2000 Transfection Reagent forms stable complexes with nucleic acid molecules, permitting efficient transfection into eukaryotic cells. Lipofectamine 2000 can be used for nuclear and cytoplasmic targets and transfects a wide variety of cell lines including CHO, HEK-293, NIH 3T3, and HeLa. Store at 4°C. See manufacturer's protocol when necessary.

7. Low efficiency of siRNA delivery with standard techniques could be related to material toxicity. The transfection efficiency of siRNAs can be improved by using transfection reagents developed for antisense oligonucleotides. These reagents can be less toxic than plasmid delivery reagents and also result in higher transfection efficiencies than conventional reagents.

8. Working rabbit sera stocks should be kept at 4°C supplemented with 0.02% sodium azide. For monoclonal antibodies such as purified IgG, diluted antibody solutions should not be stored unless detergent or carrier proteins such as goat serum or BSA are added. Diluted IgG solutions (less than 0.1 mg/ml protein) quickly adsorb, denature, and lose activity. Avoid repetitive freeze–thawing of dilute purified IgG.

9. Multiple bands on a Western blot could be indicative of a poor-quality antibody (e.g., not specific for protein of interest). It also indicates proteolytic breakdown of the antigen, excessive protein loaded per lane, a highly sensitive detection system, inefficient blocking, or the concentration of antigen is too low.

10. We have observed lower aptamer stem size between 4 and 10 bp to be optimum. The stem size >10 bp inhibits

splicing even without addition of theophylline. The upper aptamer stem size can vary between 2 and 6 bp. The pre-mRNA in which the branchpoint sequence is embedded in the upper stem displayed a strong response to theophylline-dependent control of alternative splicing.

11. We constructed pre-mRNA substrates containing the theophylline-responsive element using overlap extension PCR (*40*). This strategy removes the limitation placed by lack of restriction sites or the potential secondary effect of adding nucleotides to the pre-mRNA. In theory, two half-amplicons are generated in the first PCR reaction containing an overlap region of 20–30 nucleotides. In the second PCR, we generate the full-length product. Gel-purify each PCR product as parental DNA template can interfere with generation of the chimeric sequence. We note that success in creating the overlap extension increases when the two half-amplicons are equal in length and the same numbers of moles are used in the second PCR reaction.

12. Theophylline has clinical applications that are well documented (e.g., in the treatment of respiratory diseases with bronchospasm) but it can be toxic in high dosages (*41*). HeLa cells can tolerate dosages of theophylline from a range of 0.05 –1 mM without a significant growth defect. Our in vitro pre-mRNA splicing assays show good results with theophylline from a range of 1–2 mM.

13. Thoroughly mix reaction by repeated but gentle pipetting. We have observed inconsistent results occur from improper mixing of reverse transcriptase at this step.

Acknowledgments

We thank members of the Gaur laboratory for helpful discussions; Marieta Gencheva for valuable suggestions; and Faith Osep for administrative assistance. This work was supported in part by a Department of Defense (DOD; CDMRP) grant to RKG (BC023235), Beckman Research Institute excellence award to RKG, and NIH grant (CA 84202) to JES.

References

1. Black, D. L. (2003) Mechanisms of alternative pre-messenger RNA splicing. *Annu. Rev. Biochem.* **72**, 291–336.

2. House, A. E. and Lynch, K. W. (2008) Regulation of alternative splicing: more than just the ABCs. *J. Biol. Chem.* **283**, 1217–1221.

3. Wang, Z. and Burge, C. B. (2008) Splicing regulation: from a parts list of regulatory elements to an integrated splicing code. *RNA* **14**, 802–813.

4. Mironov, A. A., Fickett, J. W. and Gelfand, M. S. (1999) Frequent alternative

splicing of human genes. *Genome Res.* **9**, 1288–1293.

5. Johnson, J. M., et al. (2003) Genome-wide survey of human alternative pre-mRNA splicing with exon junction microarrays. *Science* **302**, 2141–2144.

6. Kan, Z., et al. (2001) Gene structure prediction and alternative splicing analysis using genomically aligned ESTs. *Genome Res.* **11**, 889–900.

7. Garcia-Blanco, M. A. (2006) Alternative splicing: therapeutic target and tool. *Prog. Mol. Subcell. Biol.* **44**, 47–64.

8. Faustino, N. A. and Cooper, T. A. (2003) Pre-mRNA splicing and human disease. *Genes Dev.* **17**, 419–437.

9. Benz, E. J., Jr. and Huang, S. C. (1997) Role of tissue specific alternative pre-mRNA splicing in the differentiation of the erythrocyte membrane. *Trans. Am. Clin. Climatol. Assoc.* **108**, 78–95.

10. Kurreck, J. (2006) siRNA Efficiency: Structure or sequence – that is the question. *J. Biomed. Biotechnol.* **2006**, 83757.

11. Leuschner, P. J., et al. (2006) Cleavage of the siRNA passenger strand during RISC assembly in human cells. *EMBO Rep.* **7**, 314–320.

12. Gaur, R. K. (2006) RNA interference: a potential therapeutic tool for silencing splice isoforms linked to human diseases. *Biotechniques* **Suppl**, 15–22.

13. Kim, D. S., et al. (2008) Ligand-induced sequestering of branchpoint sequence allows conditional control of splicing. *BMC Mol. Biol.* **9**, 23.

14. Kim, D. S., et al. (2005) An artificial riboswitch for controlling pre-mRNA splicing. *RNA* **11**, 1667–1677.

15. Kole, R., Vacek, M. and Williams, T. (2004) Modification of alternative splicing by antisense therapeutics. *Oligonucleotides* **14**, 65–74.

16. Dominski, Z. and Kole, R. (1993) Restoration of correct splicing in thalassemic pre-mRNA by antisense oligonucleotides. *Proc. Natl. Acad. Sci. USA* **90**, 8673–8677.

17. Tucker, B. J. and Breaker, R. R. (2005) Riboswitches as versatile gene control elements. *Curr. Opin. Struct. Bio.* **15**, 342–348.

18. Nudler, E. and Mironov, A. S. (2004) The riboswitch control of bacterial metabolism. *Trends Biochem. Sci.* **29**, 11–17.

19. Goguel, V., Wang, Y. and Rosbash, M. (1993) Short artificial hairpins sequester splicing signals and inhibit yeast pre-mRNA splicing. *Mol. Cell. Biol.* **13**, 6841–6848.

20. Gusti, V., Kim, D. S. and Gaur, R. K. (2008) Sequestering of the 3′ splice site in a theophylline-responsive riboswitch allows

ligand-dependent control of alternative splicing. *Oligonucleotides* **18**, 93–99.

21. Dignam, J. D., Lebovitz, R. M. and Roeder, R. G. (1983) Accurate transcription initiation by RNA polymerase II in a soluble extract from isolated mammalian nuclei. *Nucleic Acids Res.* **11**, 1475–1489.

22. Fischer, D. C., et al. (2004) Expression of splicing factors in human ovarian cancer. *Oncol. Rep.* **11**, 1085–1090.

23. Ghigna, C., et al. (2005) Cell motility is controlled by SF2/ASF through alternative splicing of the Ron protooncogene. *Mol. Cell* **20**, 881–890.

24. He, X., et al. (2004) Alternative splicing of the multidrug resistance protein 1/ATP binding cassette transporter subfamily gene in ovarian cancer creates functional splice variants and is associated with increased expression of the splicing factors PTB and SRp20. *Clin. Cancer Res.* **10**, 4652–4660.

25. Karni, R., et al. (2007) The gene encoding the splicing factor SF2/ASF is a proto-oncogene. *Nat. Struct. Mol. Biol.* **14**, 185–193.

26. Zhu, H., et al. (2005) Enhancing TRAIL-induced apoptosis by Bcl-X(L) siRNA. *Cancer Biol. Ther.* **4**, 393–397.

27. Chevinsky, A. H. (1991) CEA in tumors of other than colorectal origin. *Semin. Surg. Oncol.* **7**, 162–166.

28. Hammarstrom, S. (1999) The carcinoembryonic antigen (CEA) family: structures, suggested functions and expression in normal and malignant tissues. *Semin. Cancer Biol.* **9**, 67–81.

29. Li, W. and Cha, L. (2007) Predicting siRNA efficiency. *Cell. Mol. Life Sci.* **64**, 1785–1792.

30. Yiu, S.M., et al. (2005) Filtering of ineffective siRNAs and improved siRNA design tool. *Bioinformatics* **21**, 144–151.

31. Patzel, V., et al. (2005) Design of siRNAs producing unstructured guide-RNAs results in improved RNA interference efficiency. *Nat. Biotechnol.* **23**, 1440–1444.

32. Huesken, D., et al. (2005) Design of a genome-wide siRNA library using an artificial neural network. *Nat. Biotechnol.* **23**, 995–1001.

33. Tuschl, T. (2004) Targeting genes expressed in mammalian cells using siRNAs. *Nat. Methods* **X**, 13–17.

34. Reynolds, A., et al. (2004) Rational siRNA design for RNA interference. *Nat. Biotechnol.* **22**, 326–330.

35. Daoud, R., et al. (1999) Activity-dependent regulation of alternative splicing patterns in the rat brain. *Eur. J. Neurosci.* **11**, 788–802.

36. Venables, J. P. (2004) Aberrant and alternative splicing in cancer. *Cancer Res.* **64**, 7647–7654.

37. Tazi, J., Durand, S. and Jeanteur, P. (2005) The spliceosome: a novel multi-faceted target for therapy. *Trends Biochem. Sci.* **30**, 469–478.

38. Jenison, R. D., et al. (1994) High-resolution molecular discrimination by RNA. *Science* **263**, 1425–1429.

39. Mayeda, A. and Krainer, A. R. (1999) Mammalian in vitro splicing assays. *Methods Mol. Biol.* **118**, 315–321.

40. Ge, L. and Rudolph, P. (1997) Simultaneous introduction of multiple mutations using overlap extension PCR. *Biotechniques* **22**, 28–30.

41. Visitsunthorn, N., Udomittipong, K. and Punnakan, L. (2001) Theophylline toxicity in Thai children. *Asian Pac. J. Allergy Immunol.* **19**, 177–182.